Biomechanics of Movement
The Science of Sports, Robotics, and Rehabilitation

Thomas K. Uchida and Scott L. Delp
Illustrations by David Delp

THE MIT PRESS   CAMBRIDGE, MASSACHUSETTS   LONDON, ENGLAND

© 2020 Thomas K. Uchida and Scott L. Delp

All rights reserved. No part of this book may be reproduced in any form by any electronic or mechanical means (including photocopying, recording, or information storage and retrieval) without permission in writing from the publisher.

This book was set in Adobe Garamond by the MIT Press. Selected old style numerals were designed by Ray Larabie. Printed and bound in the United States of America.

Library of Congress Cataloging-in-Publication Data

Names: Uchida, Thomas K., author. | Delp, Scott, author. | Delp, David B., illustrator.
Title: Biomechanics of movement : the science of sports, robotics, and rehabilitation / Thomas K. Uchida and Scott L. Delp ; illustrations by David Delp.
Description: Cambridge, Massachusetts : The MIT Press, [2020] | Includes bibliographical references and index.
Identifiers: LCCN 2019050152 | ISBN 9780262044202 (hardcover)
Subjects: LCSH: Human mechanics. | Biomechanics.
Classification: LCC QP303 .D45 2020 | DDC 612.2/1—dc23
LC record available at https://lccn.loc.gov/2019050152

10 9 8 7 6 5 4 3

To my parents Carol and Marvin, and my brother Robert, for teaching me the value of swimming against the current.
—Thomas K. Uchida

To my kids, Stella and Quincy, for their love and support, and to my parents and brothers for showing me the joys of adventure and discovery.
—Scott L. Delp

To our mother who, when I was eleven, found me a great art teacher.
—David Delp

David Scott Tom

Contents

Preface xi

1 First Steps 1
  Why we study movement 4
  The Cybathlon 8
  Tools to study movement 11
  Overview of this book 15
  Language of movement 16

Part I Locomotion

2 Walking 25
  The walking gait cycle 26
  Ground reaction forces 28
  Ballistic walking model 33
  The Froude number 34
  Cost of transport 37
  Dynamic walking model 39
  Arm swing 42
  Skeletal model for gait analysis 44
  Kinematics of walking 46
  Ground reaction forces and walking speed 48
  Atypical gait 49
  Changes in walking under various conditions 52

3 Running 55
  The running gait cycle 56
  Ground reaction forces 57
  Elastic mechanisms in hopping and running 61
  Hopping robots 64
  Tuned track 66
  Elastic mechanisms to improve running shoes 69
  Leg stiffness changes with body mass 70
  Gait transitions 72
  Bipedal mass–spring model 74
  Kinematics of running 75
  Ground reaction forces and running speed 77

Part II
Production of Movement

4 ***Muscle Biology and Force*** 81
  Muscle structure 83
  The cross-bridge cycle 85
  Sarcomere structure 86
  Force–length relationship 88
  Force–velocity relationship 90
  Muscle activation 93
  Rate encoding 95
  Motor unit recruitment 96
  Electromyography 98
  Modeling muscle activation dynamics 100
  Modeling the force–length–velocity–activation relationship 102

5 ***Muscle Architecture and Dynamics*** 105
  Optimal muscle fiber length, ℓ_o^M 107
  Muscle fiber pennation angle at optimal fiber length, ϕ_o 109
  Maximum isometric muscle force, F_o^M 111
  Maximum muscle contraction velocity, v_{max}^M 112
  Tendon slack length, ℓ_s^T 114
  Measuring muscle-specific parameters 117
  Hill-type model of muscle–tendon dynamics 121
  Dimensionless curves 123
  Computing muscle force with a rigid tendon 124
  Computing muscle force with a compliant tendon 126
  Other models of muscle force generation 128

6 ***Musculoskeletal Geometry*** 133
  Muscle mechanical advantage 134
  Definition of a muscle moment arm 137
  Tendon-excursion definition of a moment arm 138
  Muscle moment arms affect muscle lengths and velocities 143
  Moment arms of multi-joint muscles 145
  Measurement and modeling of maximum joint moments 148
  Muscle architecture, moment arms, and tendon transfer surgery 152
  Moment arms of muscles with complex actions 154
  Wrapping up 156

Part III
Analysis of Movement

7 ***Quantifying Movement*** 161
  Measurement techniques 162
  Optical motion capture 166
  Unconstrained inverse kinematics 171

Transformation matrices   174
Calculating joint angles with unconstrained inverse kinematics   181
Constrained inverse kinematics   183
Kinematic model of the shoulder   186
Assessing anterior cruciate ligament injury risk   188

8 *Inverse Dynamics*   193

Measuring external forces   195
Center of pressure   197
Inverse dynamics algorithms   199
Inverse dynamics with ground reaction forces   201
Inverse dynamics without ground reaction forces   207
Verifying dynamic consistency   208
Joint moments during walking and running   209
Gait retraining to reduce knee loads and pain   212

9 *Muscle Force Optimization*   217

Biological and numerical optimizers   220
Static optimization problems solved by inspection   223
Local methods to solve static optimization problems   226
Global methods to solve static optimization problems   228
Muscle forces during walking and running   230
Estimating joint loads   238
Dynamic optimization   239
Muscle coordination during a standing long jump   242

Part IV
Muscle-Driven Locomotion

10 *Muscle-Driven Simulation*   249

Understanding muscle actions during movement is challenging   251
Creating muscle-driven simulations   254
Stage 1: Modeling musculoskeletal system dynamics   255
Stage 2: Simulating movement   259
Stage 3: Testing the accuracy of dynamic simulations   262
Stage 4: Analyzing muscle-driven simulations   269
Software for creating muscle-driven simulations   270

11 *Muscle-Driven Walking*   273

Building and testing simulations of walking   275
Muscle contributions to ground reaction forces   275
Muscle actions during the swing phase   280
Muscle actions in stiff-knee gait   282
Muscle actions over a range of walking speeds   287

Muscle actions in crouch gait   292
Heel-walking and toe-walking   297
Device-assisted walking   299

12 *Muscle-Driven Running*   305

Building and testing simulations of running   307
Muscle contributions to ground reaction forces   310
Muscle activity during running   310
Changes with running speed   312
Run-to-sprint transition   314
Muscle actions during the walk-to-run transition   315
Interaction of arm and leg dynamics   317
Swinging the legs in running   317
Foot-strike patterns   318
Device-assisted running   323
Springs to enhance running   326

13 *Moving Forward*   331

Wearable technology   332
Physical rehabilitation everywhere   335
Large-scale experiments   336
Modern statistics and machine learning   338
Modeling neuromuscular control to predict movement   340
Motivating movement   342
Open science   343
Taking the baton   347

Symbols   349
References   353
Image Credits   363
Index   365

Preface

> Start a huge, foolish project, like Noah . . .
> it makes absolutely no difference what people think of you.
> —Rumi

BIOMECHANICS HAS BEEN A POSITIVE and sustained force in my life. Playing sports dominated my life as a kid in the Delp family. I felt alive playing baseball and lacrosse, and loved to run and ski. The only two books I read before graduating from high school were a technique manual on alpine skiing and a coaches' guide to the long jump. I had dyslexia, and these rudimentary biomechanics manuals were the only books I found worth the struggle and embarrassment of reading.

I studied biomechanics in college to learn how to recover my ability to walk after injuring my hip in a ski accident. I learned everything I could about biomechanics and entered graduate school a few years later when I could walk well enough to limp around the Stanford University campus. I studied design, robotics, neuroscience, and biomechanics in graduate school. I felt I was in a good position to help people with impaired movement because I had empathy for their issues and had gained a strong background in engineering design and computer science, which was not so common in biomedical research at the time. I had the good fortune to study biomechanics with Felix Zajac, who greatly influenced my thinking.

After I finished at Stanford, I took a job as an assistant professor at Northwestern University and the Rehabilitation Institute of Chicago. As a new professor, I wanted to share my joy of biomechanics and teach principles that would enable students to understand and analyze muscle and movement. This gave rise to a course, Biomechanics of Movement, which I have taught about 30 times at Northwestern University and Stanford University. This book covers most of what I teach in this course.

By reading this book you will learn some of the most important principles of biomechanics, a field that explains how living things work using physics, and thus helps us understand and appreciate life. Biomechanics is at the center of many disciplines, including bioengineering, mechanical engineering, physical therapy, ergonomics, kinesiology, and biology, and I expect students in these fields will benefit from reading this book. I am also writing for people who are interested in neuroscience, robotics, computer science, and sports. In addition to learning principles of biomechanics, you will acquire skills for analyzing and simulating human movement; these skills will empower you to make a positive change in the world.

Many people who begin studying biomechanics fall into one of two groups: engineers who are baffled by the language of biology or biologists who are baffled by the language of engineering. Both of these languages are important. Although they are frequently taught separately, nature does not care about the boundaries that humans place between disciplines. One of the purposes of this book is to give you a good start toward proficiency in both. For that reason, I do not assume any prior knowledge of biology.

I do have to assume some familiarity with physics, as otherwise this book would be much longer than it is. Newton's laws characterize the motions of everything from planets to people. We need these laws, and other principles of mechanics, to analyze human movement, so I assume that you have a basic background in mechanics from a prior course in physics or engineering. Because mathematics is the language of mechanics, I also assume that readers have a grasp of the basic concepts of calculus, including derivatives, integrals, and differential equations. Newton's second law, the starting point for mechanics, is, after all, a differential equation.

My goal was to write an approachable book that demonstrates key concepts of biomechanics using examples that are interesting and inspiring. I did not write this book as a review of the literature. There are thousands of fascinating articles on biomechanics, and many excellent review articles; I have cited only a small fraction of these. My objective is to introduce biomechanics to people who are new to the field and enable them to explore the biomechanics literature. In addition to covering established principles, I highlight some of the big ideas that will shape the future of the field and illustrate how you can pursue these ideas to have a positive impact in science, engineering, art, and the lives of others. If you are already a biomechanics guru, I very much hope you enjoy this book and find it useful in your teaching.

I want to provide you with tools to solve practical problems. While there are no homework problems printed in the book, my colleagues and I have developed a website that has many fun and challenging problems. Solving these problems will expand your knowledge substantially. You can find these problems online at biomech.stanford.edu, which also has lecture slides, a syllabus, and other materials that you are welcome to use. You can also provide feedback at this website. Please let me know what you like and how I can improve the book. This book includes lots of data, and my collaborators and I provide much of these data at simtk.org. We established this website as a place to share biomechanical models, software, and data, and I encourage you to use and contribute to this resource.

This book contains a lot of information that has rarely been collected in one place, but it is not comprehensive. I provide few details on the role of the nervous system in coordination and control of movement, and I do not discuss how muscle changes with exercise, disuse, or age. I also include only limited discussion of the biomechanics of bone, cartilage, and ligaments. These are fascinating topics that are covered well in other books.

Writing and illustrating this book with Tom Uchida and my brother, David Delp, has been one of the most enjoyable and rewarding partnerships of my career. Much of the work was done during our "jamborees" at my kitchen table. One of our goals was to write with a single voice. Tom and I have collaborated so closely that neither of us can tell you who drafted the sentences in this book. The stories come from my experience, and we decided to write in my voice since I have taught the Biomechanics of Movement course so many times; however, Tom's fingerprints are on every page. If you appreciate the precision with which concepts of mechanics are expressed, you can thank Tom for that. If you like the pictures in the book more than the words, which most people do, you can thank my brother. He is a fantastic artist and has helped us create what we hope you will find to be a fine-looking book.

Dana Mackenzie, a mathematician and brilliant writer, dramatically improved the book by providing detailed comments and suggestions on the entire manuscript. A book should not be a chore to read, but should give the reader the same joy that it gives to the author. Dana has worked to summon that joy from chapters in which it was hidden. I am deeply grateful to him for doing this, and I think you will be too.

Silvia Blemker, a world-class biomechanist, provided early drafts of Chapters 4–6, including some great ideas for the figures and examples, and she deserves special thanks. We also thank Molly Seamans and Matthew Abbate from the MIT Press. Molly provided creative input on the book design and laid out every page. Matthew gave valuable feedback on the text and reviewed every word.

Many examples are taken from work performed in my laboratory, and my collaborators need to be recognized, which I do by naming some of them in the book and referencing others in our coauthored papers. The National Institutes of Health made this work possible, and I am deeply grateful for their support.

Other friends, colleagues, students, and mentors have made valuable contributions by reviewing early drafts of this book, coauthoring papers from which I drew inspiration, or explaining key concepts to me. I would like to thank the people pictured on the following pages for their insights and partnership. Biomechanics is a team sport, and I am grateful to be on a team with you.

I am deeply thankful for the many friends I have made in biomechanics, which continues to be a positive force in my life.

Scott L. Delp
Stanford, California

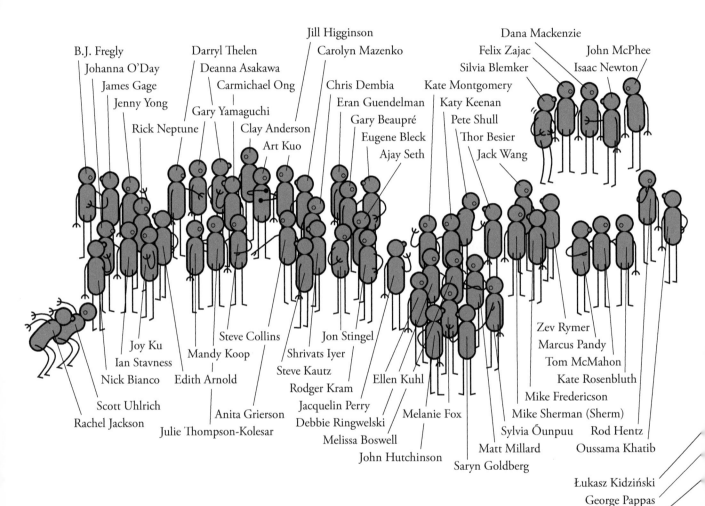

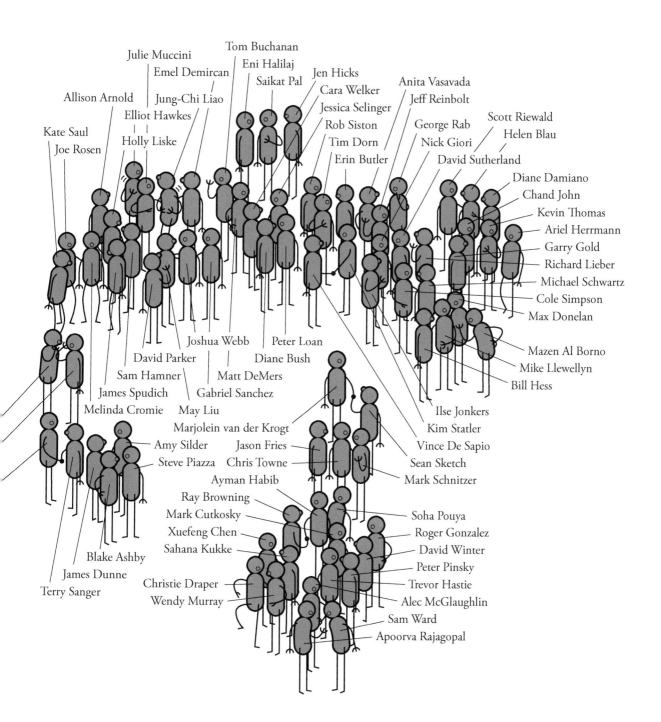

WHEN I MOVED TO CALIFORNIA with a fine arts degree, there weren't many SCULPTOR WANTED ads in the paper, so my brother Scott wrangled me some work as the Visual Information Specialist for Felix Zajac's Neuromuscular Biomechanics Lab where Scott was a grad student. Making charts and computer animations for scientists, I learned the power of clear visual communication. I got to ask really smart people the "stupid" questions: What's the main thing you want people to understand? Is a piconewton super small or super big? And can we leave that wiggly line out of the picture?

Thirty years later I call it a career. I get to ask smart people stupid questions, and Scott and Tom seem happy I do. How else could we stand each other after years making something daunting like a textbook? Here's how. We each bring a love for elegant communication, the drive to cross the finish line, and buckets of mutual respect.

David Delp
San Francisco, California

EVERY RESEARCHER KNOWS that some collaborations go more smoothly than others. I feel very fortunate to have worked with so many great collaborators. Among the first was John McPhee, my Ph.D. supervisor at the University of Waterloo, who introduced me to the world of multibody dynamics, the rigor of technical writing, and a community of inspiring people.

Working with Scott and David has been a uniquely enriching intellectual and personal experience. We all enjoyed the opportunity and I think the book is better because of it. We worked hard to fill these pages with ideas that we think are worth sharing and with illustrations that are fascinating, informative, and beautiful. We have also seasoned the text with subtly philosophical quotes and a pinch of humor to make your literary journey as palatable as possible. I hope you enjoy our creation.

Thomas K. Uchida
Ottawa, Ontario, Canada

David   Scott   Tom

1 First Steps

> It was already one in the morning; the rain pattered dismally against the panes, and my candle was nearly burnt out, when, by the glimmer of the half-extinguished light, I saw the dull yellow eye of the creature open; it breathed hard, and a convulsive motion agitated its limbs.
> —Mary Shelley, *Frankenstein* (1818)

WHEN MARY SHELLEY wrote the gothic novel that made her famous, the world was still agog over the discovery of electricity—and the shocking discovery by Luigi Galvani that electrical stimulation could cause a dead frog's muscles to twitch. It was a short leap for Shelley to imagine an entire being brought to life by a jolt of electricity.

Today, two centuries after Galvani's demonstration, technology for electrical stimulation of muscles has a profoundly positive impact on life. Electrical defibrillators are installed in every airport, school, and hospital, and have restarted the hearts of thousands of people who would have died without a powerful shock to their heart muscle. A more sophisticated technology, called functional electrical stimulation, has been developed to reanimate the muscles of people with various forms of paralysis and thus improve their quality of life.

In functional electrical stimulation, electrodes are either affixed to the skin or implanted in paralyzed muscles near the nerves that innervate them. Electrical impulses are transmitted through the electrodes to nerves which, in turn, cause the associated muscles to contract and generate force. Electrical stimulation of muscles has enabled people with paralysis to stand and walk when the trunk and leg muscles are properly activated and appropriately coordinated. In 2016, several people who had injuries to their spinal cord that resulted in paralysis from the chest down competed in a bike race as part of a new international competition called the Cybathlon. The winner pedaled 750 meters in less than 3 minutes.

There is much more to functional electrical stimulation than merely delivering a jolt of electricity to the nerves and muscles. In able-bodied people, the nervous system coordinates many muscles to enable us to walk, run, or ride a bike, much like a conductor coordinating the musicians of an orchestra. As the winner of the Cybathlon competition pedaled around the track, up to 24 channels of precisely timed stimuli could be delivered to coordinate the forces generated by his otherwise paralyzed muscles. Even with this level of complexity, the results were less than perfect. We still have much to learn about how our muscles work together to create the music of movement.

In this book, we will explore this magnificent symphony. We will examine the movement of humans and animals through the lens of mechanics to understand the *biomechanics of movement*. We will use simple conceptual models to answer questions like why walking and running are efficient forms of locomotion, why astronauts adopt a bounding gait on the moon, and how running tracks and shoes can increase performance while reducing injuries. We will examine the structure of a muscle down to its microscopic force-generating motors and see precisely how electrical stimulation causes muscles to contract. We will describe sophisticated computational tools for generating simulations of movement, which in turn allow us to estimate the muscle forces responsible for producing this movement. Such simulations show us how we coordinate our muscles when walking, running, or cycling. Indeed, muscle-driven simulations were a valuable tool for the gold medal team in the Cybathlon.

Throughout the book, we emphasize established theories to provide a foundation for understanding movement biomechanics and include innovations in areas like computer simulation, mobile motion monitoring, and wearable robotics that build on these fundamentals. Many recent advances were foreseen by science fiction and provide glimpses of new technologies we imagine for the future. Some of these visions may seem as fantastical as

FIGURE 1.1
Page from Leonardo da Vinci's notebook showing his concept of muscle lines of force. Image courtesy of the Royal Collection Trust.

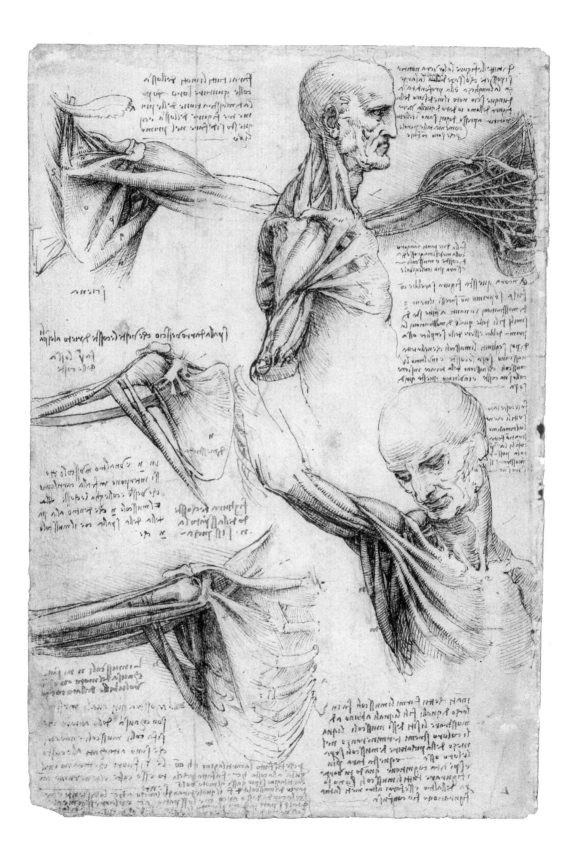

awakening Frankenstein's Creature; others are well within sight. Either way, we are just opening our eyes to the potential of this exciting field.

Why we study movement

Movement is fascinating and fundamental to life. Our bodies are amazingly versatile, exhibiting both strength and dexterity. Movement is vital for maintaining physical and psychological health. Regular physical activity helps prevent heart disease, cancer, osteoporosis, obesity, diabetes, depression, anxiety, and other serious illnesses, yet less than half of the world's population is sufficiently active to maintain their physical and mental health. Physical activity is potent and inexpensive medicine that can have profound health benefits, even in small doses. In studying the biomechanics of movement, we seek to understand the biological structures and processes involved in producing movement, and to apply this knowledge to improve mobility, physical activity, and health.

The field of movement biomechanics has a rich history. As with many human pursuits, advances have been driven by a desire to improve our lives coupled with an innate curiosity about our environment and ourselves. Aristotle wrote the first book that examined the general principles of animal locomotion, aptly titled *On the Motion of Animals*, around 350 BCE. He and other ancient Greeks thought that muscles contract when they are inflated by *pneuma*, the "breath of life" flowing through our nerves.

Countless scholars have since advanced the field, some of whom are featured in this book. Among the pioneers of biomechanics are Leonardo da Vinci (1452–1519), who produced hundreds of detailed anatomical drawings and who studied the mechanical function of the musculoskeletal system (Figure 1.1). Giovanni Borelli (1608–1679) was the first to apply the laws of mechanics to relate the force exerted by a muscle to the moment it generates about a joint (Figure 1.2). Borelli, however, still held to the classical view that muscles contract by a pneumatic, inflationary process. This theory was refuted by Nicolas Steno (1638–1686), a rival of

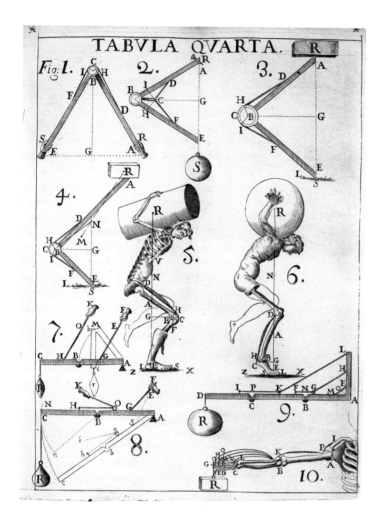

FIGURE 1.2
Static analysis of human muscles and joints by Giovanni Borelli, often described as the father of biomechanics. Image from *De motu animalium*.

Borelli in the Florentine court, who pointed out that muscle retains a constant volume while contracting—a paradox that we will examine in Chapter 4. Later, Luigi Galvani (1737–1798) discovered the previously unsuspected ability of electrical signals to generate muscular contraction, essentially founding our modern field of electrophysiology. Finally, I cannot fail to mention Eadweard Muybridge (1830–1904), who performed early photographic studies of human and animal movement about a kilometer from my home on what is now the Stanford University campus. Muybridge's photos are fascinating to look at even today, and they were a landmark in

the history of both film and biomechanics (Figure 1.3). Indeed, the technology of filmmaking and the science of biomechanics still advance hand in hand.

Today, biomechanics is a rapidly growing, multidisciplinary field involving collaborations between individuals from many areas of science and engineering. Biologists use insights from biomechanics to understand the relationship between form and function in animals: for example, how a lizard leaps and grabs onto a wall (Figure 1.4) or whether *Tyrannosaurus rex* was capable of running (Chapter 3). Neuroscientists study how the brain coordinates our muscles during

FIGURE 1.3
Sequence of photographs from Eadweard Muybridge, a pioneer in motion capture. Images courtesy of Stanford University.

FIGURE 1.4
The red-headed agama uses its tail to control its body orientation during flight. The lizard prepares to grab the vertical wall at right by rotating its tail clockwise, which orients its body vertically through conservation of angular momentum. Image courtesy of Robert Full.

movement and how these neural circuits are disrupted in cases of injury and disease. Surgeons can use biomechanical models to determine whether a patient with cerebral palsy will benefit from tendon lengthening surgery. Roboticists are inventing prostheses that are increasingly sophisticated (Figure 1.5) and bipedal robots that can perform complex tasks in dangerous environments. Sports scientists analyze athletic movements, such as the "Fosbury flop" high-jump technique (Chapter 9), to learn how to increase performance and prevent injuries. Computer scientists and biomechanical engineers develop new algorithms and software tools to simulate movements and gain insights from these simulations.

Biomechanics plays a role even outside the world of science. Filmmakers use biomechanics and motion capture techniques to create computer-generated imagery for games and movies that range in style from fantastical to photorealistic (Figure 1.6). The results marry beauty with scientific accuracy in a way that would have been hard to imagine a generation or two ago.

Collectively, these efforts have benefited our lives immensely. We now have recommendations for the daily physical activity levels we require to stay healthy, and we have access to tools that help us achieve our fitness goals. Assembly lines, office furniture, and many consumer products are designed ergonomically to improve comfort and prevent injury. Products and procedures have been designed to replace hip and knee joints, reducing pain and restoring function to millions of people with osteoarthritis. Powered exoskeletons are revolutionizing post-stroke rehabilitation and can enable locomotion in individuals with paralysis. Movement disorders such as Parkinson's disease have been successfully treated using deep brain stimulation, in which implanted electrodes deliver electrical impulses to specific regions of the brain. Epidural stimulation, a technology that activates neural circuits in the spinal cord, has recently shown promise for helping to restore voluntary movement to individuals with spinal cord injuries. Athletes are benefiting from equipment and training programs designed to reduce injury risk while enhancing performance (Figure 1.7).

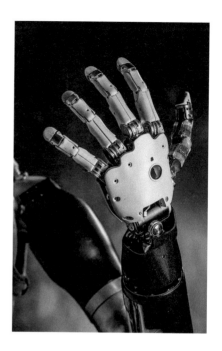

FIGURE 1.5
Highly functional prosthetic hands are now within reach. Image courtesy of Johns Hopkins University.

FIGURE 1.6
Image from the movie *Avatar* and the performance artist Jenn Stafford, whose motions were recorded with motion capture to produce computer-generated imagery. Images courtesy of Twentieth Century Fox.

The Cybathlon

The impact of biomechanics can perhaps be best appreciated by looking more closely at a compelling example—the Cybathlon. In 2016, Robert Riener, a professor at ETH Zurich (the Swiss Federal Institute of Technology), organized the first Cybathlon, an innovative merger of science and athletics. Unlike the Paralympics, it had events that focused on everyday tasks. For example, people with prosthetic arms competed at carrying objects, slicing bread, opening a jar, and hanging laundry. People with prosthetic legs competed at climbing ramps and ascending a staircase while balancing a cup on a saucer. To emphasize that the competition was about the interface between humans and technology, the competitors were called pilots rather than athletes.

The most traditional sporting event in this untraditional competition was the bicycle race for people with paralysis of the legs. The gold medal went to a team from Case Western Reserve University, under the scientific leadership of Ronald Triolo. The team's pilot was Mark Muhn (Figure 1.8), who had injured his spinal cord in a ski accident in 2008 and was paralyzed from the chest down.

The team at Case Western had been doing research on functional electrical stimulation for forty years and developed the first implanted neuromuscular stimulators. Triolo believed that their experience with implanted electrodes would be a winning advantage, because the other teams would be using electrodes that transmit electrical signals through the skin. Implanted electrodes allow stimulation to be more specific about which muscles are activated, and the resulting muscle contractions can be stronger.

The team maximized their chances of winning in several ways (McDaniel et al., 2017). They purchased a recumbent bike and stripped it down, removing all nonessential weight. They put their pilots (originally five, of whom two were selected to go to Zurich) through an intensive training program. Of most interest to us is their use of biomechanical models (Figure 1.9).

Triolo says that the models, similar to the ones we describe later in this book, were helpful in two ways: first, to eliminate "dead ends" (ideas that wouldn't work), and second, to optimize the electrical stimulation patterns provided to the pilot. Every contraction of every leg muscle was controlled by an external control unit, a box strapped to the pilot's waist. For safety, the pilot controlled whether the box was switched on or off.

To obtain the stimulation pattern, the team took biomechanical models of cycling motion from the literature and customized them to the characteristics of each pilot. Customization was important because the properties of muscle change after paralysis, and the number of muscles that can be activated with stimulation is small, so the activation pattern that would work for an able-bodied cyclist might not work for stimulation-driven pedaling in someone with a spinal cord injury. Musa Audu led the effort to model and customize the stimulation patterns for each pilot, and employed techniques described in Chapter 10 to estimate the timing and intensity of the activation of each muscle. It is worth noting that the stimulation pattern never changed during the race. As the pilot's muscles fatigued, his legs would keep pumping at the same rate, but they would not be able to push as hard so the pilot had to shift gears to keep the bike going.

FIGURE 1.7
Biomechanical studies help create sports equipment that maximizes athletic performance. The hinge between the boot and blade of this skate prolongs contact between the blade and the ice, increasing the duration of propulsive force to increase speed. Image courtesy of McSmit.

FIGURE 1.8
Mark Muhn competing at the Cybathlon. Image courtesy of Paul and Gabrielle Marasco.

In fact, the pilots tired rapidly because functional electrical stimulation does not recruit muscle fibers in the same way that natural signals from the brain do. We will review this phenomenon in Chapter 4, but for now it suffices to say that most naturally occurring movements are generated by recruiting "slow-twitch" muscle fibers first (which are relatively small and fatigue-resistant), followed by "fast-twitch" fibers (which produce larger forces but tire quickly). However, electrical stimulation recruits muscle fibers in the opposite order. As a result of this "reverse recruitment," the pilots could scarcely keep the bike going for more than a minute when they began their training.

But something unexpected came to the rescue—the pilots loved the outdoor exercise that race-ready bikes provided, compared to the humdrum labor of stationary bikes. And exercise training can make the paralyzed muscles stronger. With the pilots motivated to improve, after 5 months of training they were able to keep cycling for a 3-minute race. In future years, Triolo says he would like to come up with a technological solution for the recruitment reversal issue, but in 2016 there was only one solution: to "exercise the heck out of our pilots." And it worked! In Zurich, Muhn finished the 750-meter course in 2 minutes and 58 seconds, and won the gold medal.

The Cybathlon experience changed Muhn's life, and perhaps more surprisingly, it changed Triolo's research. Previously, Triolo

FIGURE 1.9
Example of a muscle-driven biomechanical model used to tune muscle excitation patterns to produce cycling.

says, their approach to rehabilitation had been task-oriented. They focused on getting their volunteers to stand and walk or perform activities of daily living, like dressing and cooking. But when their participants rode their bikes outside, and started having fun while exercising, it changed everything. They exercised more, and it paid huge dividends for all aspects of their rehabilitation, including boosting their self-esteem. While one pilot was riding his bike on a public trail, another cyclist caught up to him and said, "Nice wheels." It was practically the first time that a stranger saw him not as a person with a disability, but as a person with a cool bike.

Tools to study movement

One of my goals in this book is to familiarize you with the muscle-driven biomechanical models that my team and others have developed. It is important to realize that these models are informed by experimental data. Let's take a look at the types of data we collect.

A common technique for analyzing movement is to record video of individuals in research labs and clinics. Many of these recordings are obtained with infrared cameras that track markers affixed to the skin, similar to the technology used in filmmaking (Figure 1.10). Recently, motion capture techniques that do not require markers have become more popular. Video-based systems are widely available, but the

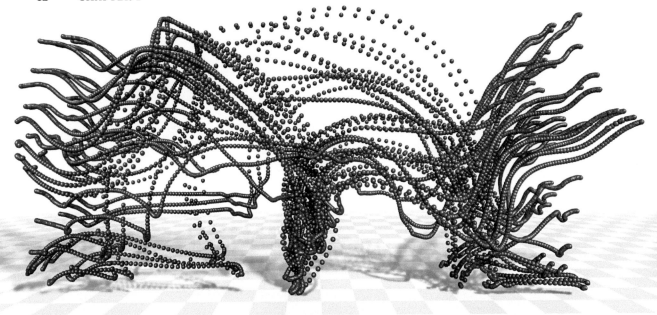

FIGURE 1.10
Trajectories of markers affixed to the skin, captured during a front handspring (2.9-second duration). The subject was initially standing (far left), then skipped forward, flipped over his hands planted on the floor (center), landed on his feet, and took one hop to regain his balance (far right). Spheres that are further apart indicate that the marker was moving faster. Data from ACCAD (2018).

prevalence of these systems has been eclipsed by inertial measurement units (IMUs), which are now integrated into smartphones, wearable activity monitors, and clothing. IMUs enable collection of movement data (velocities and accelerations) in natural environments and over long durations, which can be valuable for monitoring disease progression and planning treatments. The availability of IMUs in low-cost activity monitors is also enabling large-scale studies of millions of individuals across the globe (Figure 1.11).

When performed in the lab or clinic, experiments may involve several pieces of specialized equipment beyond a motion capture system. We often use force plates to measure the forces between the feet and the ground. In studies of walking and running, it is convenient to use a treadmill so that the subject remains within the volume that can be accurately measured by the motion capture system. Some treadmills are instrumented to measure ground reaction forces as well. Electromyography is used to measure the timing and intensity of activity in various muscles. We may monitor a runner's breathing, measuring the amounts of oxygen consumed and carbon dioxide produced, to estimate the metabolic energy required to run. The proliferation of imaging technologies, such as

FIGURE 1.11
Smartphone data from over 68 million days of activity by 717,527 individuals reveal variability in physical activity across 111 countries. Image adapted from Althoff et al. (2017), the world's largest survey of physical activity.

magnetic resonance imaging and fluoroscopic imaging, has enabled us to see inside an animal or human body in motion, providing a powerful tool to visualize and measure movement.

Conceptual models can be powerful analytical tools, as we will see throughout the book. Chapters 2 and 3, for instance, show how a simple pendulum model gives us a reasonable facsimile of walking, and a mass–spring model can give us important insights about running. A simple model that gets the job done is almost always preferable to a more complicated version, the latter providing similar utility but being more difficult to construct and understand. Of course, not all complex phenomena can be represented with simple mechanical models. As we will see in Chapters 11 and 12, muscle-driven simulations are powerful tools that can fill in many of the details that are missing from the simple models of Chapters 2 and 3.

Computer simulations that represent the dynamics of movement complement experiments by calculating quantities that cannot be directly measured and by predicting movement in hypothetical scenarios. Simulations can be useful for understanding injuries, for example, which are difficult to study experimentally. We may also estimate the muscle forces that are responsible for an observed

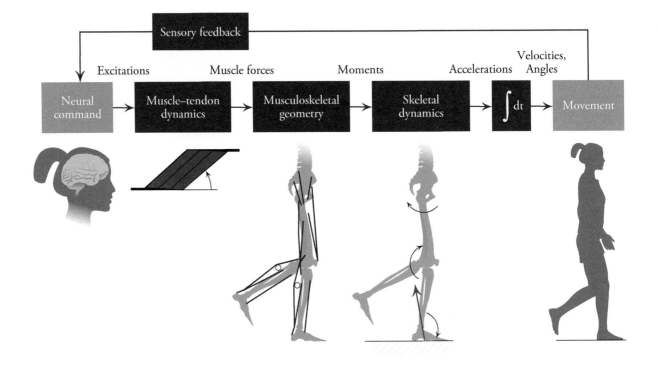

FIGURE 1.12
Elements of a typical *forward dynamic* simulation. Movement arises from a complex orchestration of the neural, muscular, skeletal, and sensory systems. Computational models of these systems enable us to predict and analyze human and animal movement.

motion, as well as joint loads, tendon strains, and other quantities that cannot be measured. In a *forward dynamic* simulation, we prescribe a pattern of neural activation of one or several muscles, and then predict the resulting motion of a musculoskeletal model (Figure 1.12). Experimental data are required to develop and test the mathematical models of musculoskeletal dynamics that are used in simulations of movement, and to evaluate the degree to which simulations reflect reality.

The reverse process is often used when we have measurements of the motion of a test subject and we want to convert these data into meaningful insights, such as what forces must have been generated by the muscles to produce the measured motions. For this we need an *inverse dynamic* analysis, a common analytical strategy that combines experimental data with musculoskeletal models. The first step is to use a biomechanical model of the body to convert measurements of marker positions, as shown in Figure 1.10, into joint angles through a process called inverse kinematics (Figure 1.13).

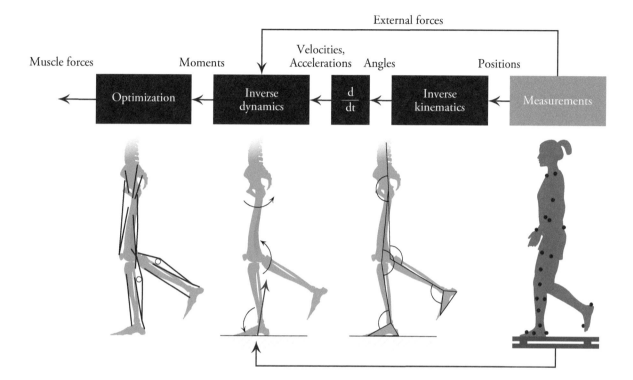

The joint angles are differentiated with respect to time to estimate joint angular velocities and accelerations, which are used with measurements of the external forces applied to the body to estimate joint moments. A biomechanical model is then used with an optimization algorithm to estimate muscle forces. We call this strategy an inverse analysis because measurements of motion are used to deduce what forces must have been present to produce those motions.

FIGURE 1.13
Elements of a typical *inverse dynamic* analysis. The analysis begins with measurements of marker trajectories and external forces (right) and uses a biomechanical model to estimate the angles, velocities, and accelerations of the body segments and joints. An inverse dynamic model and optimization procedure produce estimates of joint moments and muscle forces.

Overview of this book

In the chapters that follow, we begin by studying the two common forms of human locomotion, walking and running, using simple conceptual models. We then explore the production of movement by examining the biology and architecture of skeletal muscle, its dynamic interaction with tendon, and how muscles generate the forces that animate the skeleton. Next, we examine the models and algorithms used to analyze movement. We demonstrate methods to compute joint angles, joint moments, and individual muscle forces from motion capture data and biomechanical models. Finally, we synthesize these concepts to study the roles of muscles during walking and running, and conclude with some ideas about where I think the field is headed. As shown in Figure 1.14, the material has been arranged into four parts, and I encourage readers to think of the book in this way. Thus, Chapter 4 (the beginning of Part II) is not exactly a sequel to Chapter 3, but Chapter 5 definitely picks up where Chapter 4 ends. The material will have more meaning for you if you keep this in mind.

While we focus primarily on human locomotion, the fundamental concepts we describe can be used to understand the motion of animals and robots as well. The material covered here will allow you to understand the fantastic and broad scientific literature that provides detailed analyses of more topics than could be covered in a single book.

FIGURE 1.14
Organization of this book.

Language of movement

I would like to explain what you should know to get the most out of this book. Tom and I have written the book for readers who are familiar with certain engineering fundamentals. Mathematics and mechanics provide a precise framework for analyzing movement, and we assume a basic understanding of vectors and matrices. We further assume that the reader is familiar with free-body diagrams, deriving equations of motion, and solving these equations for simple systems. If these topics are unfamiliar, you should be prepared to spend some extra time with them when they come up.

On the biological side, a background in human anatomy and physiology is helpful but not required. For those readers who are not familiar with anatomy, the figures that follow illustrate the terminology that will be used throughout this book. These terms include the anatomical planes and directions (Figure 1.15), joint motions (Figures 1.16 and 1.17), and major bones and muscles (Figures 1.18 and 1.19). This nomenclature may look intimidating

FIRST STEPS 17

Part I
Locomotion

Chapters 2 3

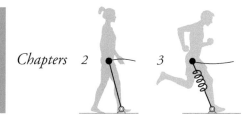

Part II
Production of Movement

Chapters 4 5 6

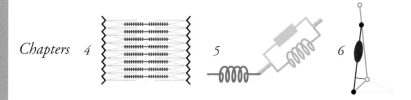

Part III
Analysis of Movement

Chapters 7 8 9

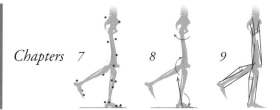

Part IV
Muscle-Driven Locomotion

Chapters 10 11 12

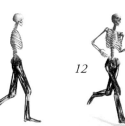

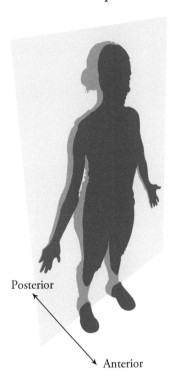
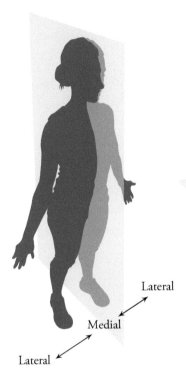
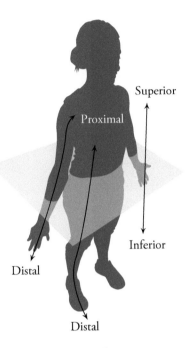

FIGURE 1.15 (ABOVE)
Anatomical planes and directions in a human.

FIGURE 1.16 (OPPOSITE)
Motions of the shoulder, elbow, pelvis, and hip in the frontal plane (left), sagittal plane (center), and transverse plane (right).

FIGURE 1.17 (LEFT)
Motions of the knee and ankle in the sagittal plane.

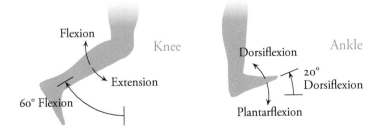

FIRST STEPS

	Frontal plane	Sagittal plane	Transverse plane
Shoulder	Abduction / Adduction — 40° Abduction		
Elbow		Flexion / Extension — 80° Flexion	
Pelvis	List	Tilt — Posterior / Anterior	Rotation
Hip	Abduction / Adduction — 20° Abduction	Flexion / Extension — 30° Flexion	Internal rotation / External rotation — 80° External rotation

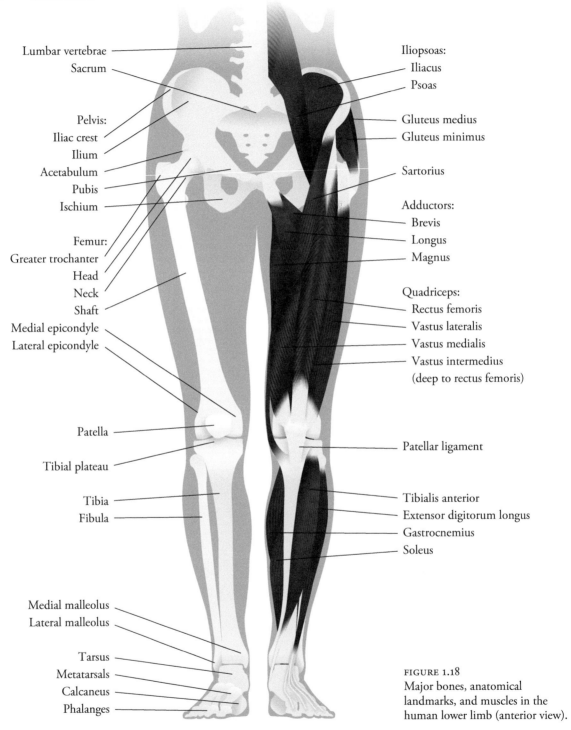

FIGURE 1.18
Major bones, anatomical landmarks, and muscles in the human lower limb (anterior view).

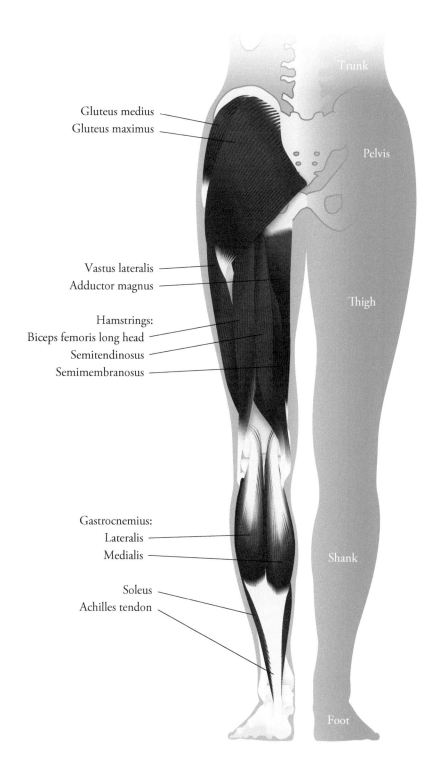

FIGURE 1.19
Body segments and major muscles in the human lower limb (posterior view).

at first, but a few minutes spent studying these figures and learning these words now will serve you well.

This book is only the start of what I hope will be a continuing journey for you. Our dream is that you will build on the material assembled here, adding your unique spark of creativity to discover something about nature and invent something that enriches the lives of others.

Part I
Locomotion

2 Walking

With each step, you fall forward slightly
And then catch yourself from falling
Over and over, you're falling
And then catching yourself from falling
—Laurie Anderson

"I WAS STROLLING ON THE MOON ONE DAY," sang astronaut Harrison Schmitt cheerfully, as he set out on the first moonwalk of the last Apollo mission. "In the merry, merry month of May," chimed in Commander Gene Cernan. Tufts of moondust sprayed from their boots as they ... did what, exactly? hopped? pranced? stumbled? ... across the surface of the moon.

Whatever it was, it bore little resemblance to what we normally think of as walking. Schmitt loped along and swayed from side to side like a toddler. Cernan's gait looked like a child riding a broomstick and pretending it is a horse. It was as if the two highly trained astronauts had forgotten the most basic human skill—walking—and had to learn a new way to move (Figure 2.1).

FIGURE 2.1
Astronauts rarely "walk" on the surface of the moon, preferring a hopping gait in lunar gravity. Images courtesy of NASA.

In this chapter, we'll see why the seemingly simple task of walking, which human evolution has perfected for Earth's gravity, becomes so difficult on the moon. The astronauts' bizarre gaits may help us to appreciate more fully the carefully orchestrated sequence of events that occur when we walk, and the pivotal role that gravity plays.

Another way to appreciate the feat of walking is to build a machine that does it. Ingenious experiments by Tad McGeer have shown that an unpowered mechanism with human-like proportions can walk down a sloped surface. No brain, spinal cord, or muscles are needed for these mechanisms to walk, just a little push in the right direction. This observation suggests that we can learn a lot about walking using simple mechanical models, which we will demonstrate in this chapter.

Before beginning our analyses, however, it is useful to examine the walking gait cycle and some of the basic physics involved in locomotion. The next section will give you a little push in the right direction.

The walking gait cycle

Humans have two common gaits: walking and running. We are all familiar with the stereotypical cyclical patterns our body segments undergo when we walk, but it is useful to characterize these qualitative observations more formally. A walking gait cycle is

FIGURE 2.2
The walking gait cycle and its constituent events (e.g., foot contact) and phases (e.g., double support).

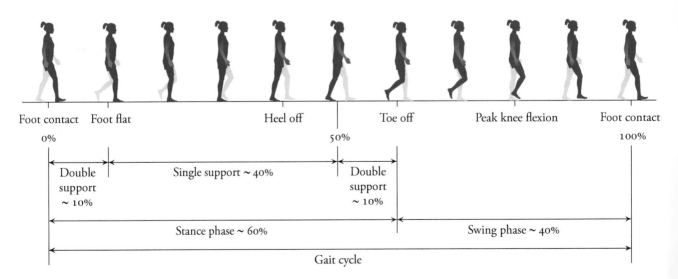

delimited by two consecutive foot contact events on the same leg, with foot contact of the opposite leg typically occurring halfway through (Figure 2.2). Each leg has a stance phase, when the foot is in contact with the ground, and a swing phase, when the foot is off the ground. Stance begins at foot contact, ends at toe off, and accounts for about 60 percent of the gait cycle in walking for a given leg; its remaining time is spent in swing. Because the stance phase is longer than the swing phase, there are periods in each walking gait cycle when both feet are in contact with the ground, which we call double-support phases. We refer to the intervals when only one foot is in contact with the ground as single-support phases.

The step length is the distance along the line of progression between the same point on two consecutive footprints (Figure 2.3). The distance traveled in two consecutive steps, or the distance covered in one gait cycle, is called the stride length. The rate at which foot contact events occur (equivalently, the reciprocal of step duration) is called the step frequency or cadence; the rate at which strides are taken is called the stride frequency. Walking speed can be computed as the product of stride length and stride frequency or, equivalently, step length and cadence:

$$\begin{aligned} \text{speed} &= \text{stride length} \times \text{stride frequency} \\ &= \text{step length} \times \text{cadence} \end{aligned} \quad (2.1)$$

The speed at which individuals typically walk varies with height, fitness, and other factors. Healthy adults usually choose to walk at speeds of about 1.2–1.4 m/s on level ground, with a cadence of 2 steps/s. A typical step length is around 0.6–0.7 m.

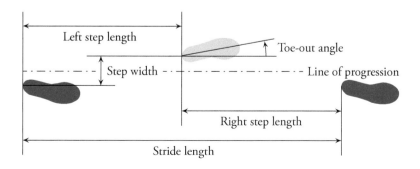

FIGURE 2.3
Measurements of gait in the horizontal (ground) plane.

Two additional metrics of note are measured in the horizontal plane (Figure 2.3). Step width is the distance between the middle of the heel at foot flat and the same point on the opposite leg, measured perpendicular to the line of progression. Step width is approximately 10 percent of leg length in healthy adults. If leg length is about 1 m, step width will be about 10 cm, but it is larger in toddlers who are learning to walk and in some people who have poor balance. Foot progression angle is the angle between the line of progression and the line joining the midpoint of the heel and the second toe (i.e., the long axis of the foot). Positive and negative foot progression angles are referred to as toe-out and toe-in angles, respectively. A small amount of toe out, about 10 degrees, is typical for adults, but this value can vary among individuals with musculoskeletal or neurological impairments. For example, individuals with cerebellar ataxia, which causes disturbances in balance, may walk with greater toe out and step width to reduce the risk of falling.

Ground reaction forces

We study walking by measuring the forces between the feet and the floor (Figure 2.4). We will learn more about muscle coordination during walking in Chapter 11; for now, it is sufficient to know that muscles produce movement by generating forces. The "action" forces generated by muscles result in "reaction" forces applied by the ground to the foot. Ground reaction forces can be measured during walking with a force plate, an instrument that measures the forces in the vertical, anteroposterior (fore–aft), and mediolateral (side-to-side) directions as a person walks over the plate. Ground reaction forces are important because they provide a measure of how the body's center of mass is accelerating at each instant in time. We can use Newton's second law to relate the ground reaction forces and other external forces to the acceleration of the center of mass (COM) of a body:

$$F_{external} - mg = ma_{COM} \qquad (2.2)$$

where $F_{external}$ is the sum of all external forces applied to the body, m is the total mass of the body, g is gravitational acceleration, and

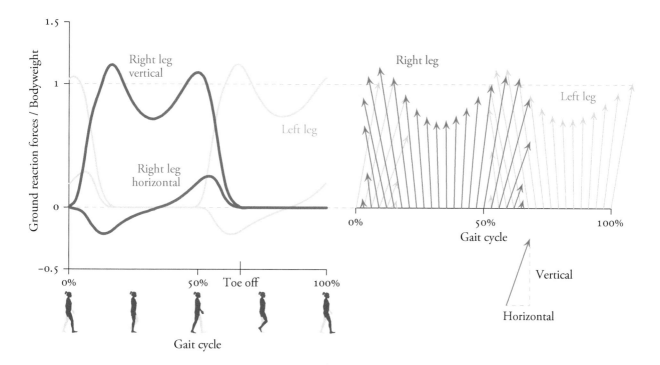

a_{COM} is the acceleration of the center of mass. It's worth taking a moment to think about the significance of the words "sum" and "all" in the previous sentence. When both feet are contacting the ground, we must add the forces applied to each foot. Because Equation 2.2 is a vector sum, directions matter. The vertical components of the forces under both feet support the body's weight. But, as you can see in Figure 2.5, the fore-to-aft force on the lead foot tends to cancel the aft-to-fore force on the trailing foot; that is, the lead foot serves as a brake that prevents us from going faster and faster, and the trailing foot provides propulsive force. If no external forces are applied (i.e., $F_{external} = 0$), then the body is in freefall and its center of mass will accelerate toward the ground at $g = 9.81 \text{ m/s}^2$.

Recordings of ground reaction forces during walking show several interesting features (Figure 2.6). The vertical component of the ground reaction force rises rapidly after foot contact and reaches a force greater than body weight at about 10 percent of the gait cycle.

FIGURE 2.4
Representative ground reaction forces during walking at 1.55 m/s. Vertical and horizontal (fore–aft) components (left) and schematic of total vector (right) are shown over the gait cycle. Positive horizontal forces are directed forward. Side-to-side forces, which are smaller, are not shown. Data from Dembia et al. (2017).

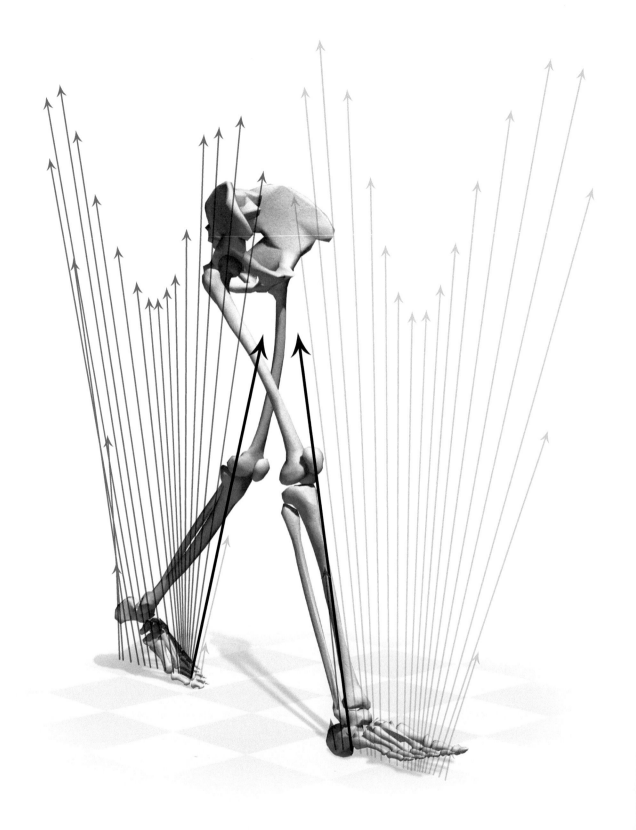

The vertical force falls below body weight at midstance and then rises again above body weight during push off. The vertical force then falls to zero because, after toe off, the foot is no longer in contact with the ground. On average, the total vertical ground reaction force is 1 bodyweight. Notice that, according to Equation 2.2, the center of mass has no net vertical acceleration when the vertical component of the ground reaction force is equal to body weight, exactly balancing the downward force due to gravity.

The horizontal ground reaction force is directed toward the back of the body, decelerating the center of mass, during the first half of stance and is directed toward the front of the body thereafter. In addition to the fore–aft component shown in Figure 2.4, the horizontal ground reaction force has a mediolateral component, which is small but important for controlling side-to-side balance. Although at any given instant the center of mass may be accelerating, the average fore–aft acceleration over a few steps of steady-speed walking is zero. Average fore–aft acceleration will be nonzero when changing forward speed. In the vertical direction, nonzero average accelerations occur when changing slope, such as when walking from level ground onto a ramp.

Bear in mind that during part of the gait cycle both feet are touching the ground and the two forces are summed to obtain the net upward force on the body (Figure 2.5). One ground reaction force is directed upward and forward, while the other is directed upward and backward. The net ground reaction force is mostly upward, with a small forward component at the beginning of the

FIGURE 2.5 (OPPOSITE) Ground reaction forces measured by two force plates during walking at 1.55 m/s. The two sets of arrows illustrate the force applied to each foot over time. As indicated by the black arrows, the fore–aft components of the ground reaction forces point in opposite directions during double support.

FIGURE 2.6 Representative ground reaction force applied to the foot during walking at 1.55 m/s. The force vectors shown here (in space) are the same as those shown in Figure 2.4 (over time).

Heel strike

Loading

Midstance

Push off

Toe off

double-support phase and a small backward component at toe off. These forces counteract the downward force of gravity and help you modulate your walking speed.

Force plate recordings can be divided by the total body mass to estimate the acceleration of the center of mass (a_{COM} in Equation 2.2). This acceleration can be integrated once to estimate the velocity of the center of mass (v_{COM}) and twice to estimate its position (r_{COM}). From these quantities, the kinetic energy of the center of mass in the forward direction (E_{kf}) can be estimated as

$$E_{kf} = \frac{1}{2} m v_{COM,f}^2 = \frac{1}{2} m \left(\int a_{COM,f}(t) \, dt \right)^2 \quad (2.3)$$

where the "f" subscript denotes the forward component. (Please note that the fluctuation of the velocity in the vertical direction is very small and thus we are ignoring it here.) The integral sign in Equation 2.3 reminds us that the effect of acceleration is cumulative and has to be integrated over time if we want to know the velocity or kinetic energy. The forward kinetic energy is lowest at midstance because the horizontal ground reaction force is directed backward, decelerating the center of mass, during the first half of stance. The gravitational potential energy can be estimated as

$$E_{pg} = m g r_{COM,v} = m g \iint a_{COM,v}(t) \, d^2 t \quad (2.4)$$

where the "v" subscript denotes the vertical component. The gravitational potential energy is highest at midstance as the body vaults over the stance limb (Figure 2.7). If we plot the forward kinetic energy and the gravitational potential energy over the gait cycle, we see that they are out of phase (Figure 2.8): forward kinetic energy is at its peak when gravitational potential energy is near its minimum and vice versa. As a result, the total energy is nearly constant. One of the means by which we conserve energy during walking is by trading gravitational potential energy for forward kinetic energy or forward speed, similar to what occurs in a pendulum oscillating under the influence of gravity. This observation suggests that gravity has a lot to do with walking.

FIGURE 2.7
Vertical motion of the center of mass over one walking gait cycle.

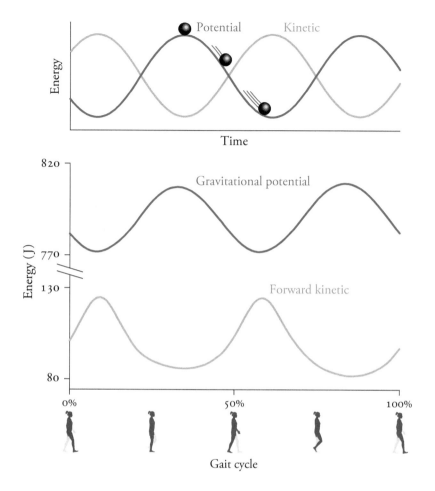

FIGURE 2.8
Representative gravitational potential and forward kinetic energies during walking at 1.55 m/s (bottom) resemble the exchange of potential and kinetic energies in a ball rolling up and down hills (top). During walking, fluctuations of kinetic energy in the vertical direction are small and not shown. Human data adapted from Dembia et al. (2017).

Ballistic walking model

In 1980, Thomas McMahon and Simon Mochon developed a mathematical model for walking based on the assumptions that the body exchanges potential for kinetic energy and minimizes the role of muscles during the swing phase. The model is very simple, consisting of only three rigid links and three revolute (pin) joints—and, notably, no muscles or motors (Figure 2.9). The stance limb is represented by an inverted pendulum that is pinned to the ground at the ankle, allowing the limb to rotate about the ankle in the sagittal plane; the stance-limb knee is assumed to remain locked in a fully extended position. The swing limb is modeled as a double pendulum, with the thigh and shank segments pin-connected at the knee. The two legs are pinned together at the hip and have realistic mass distributions. The head, arms, and trunk are ignored, except for their mass, which is concentrated at the hip.

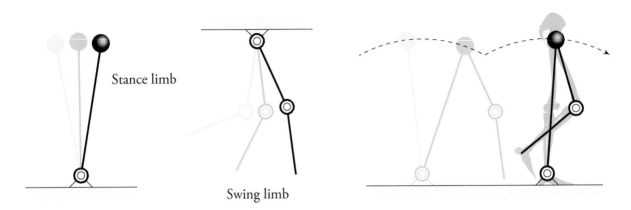

Experimental recordings of muscle activity have revealed that the muscles in the swing limb are relatively inactive during normal-speed walking, except at the beginning and end of the swing phase. Accordingly, the ballistic walking model assumes that the muscles establish the positions and velocities of the limb segments necessary for the model to initiate and evolve through the swing phase entirely passively, acted upon only by gravity (like a projectile, hence "ballistic"). When the muscles are inactive, the swing limb is assumed to behave like an unforced double pendulum. If just the right initial conditions are provided to the model at toe off, the toe of the swing limb will clear the ground at mid-swing and the knee will reach full extension at foot contact.

FIGURE 2.9
The ballistic walking model described by Mochon and McMahon (1980) is simple yet analytically powerful. The stance limb is represented by an inverted pendulum (left); the swing limb, by a double pendulum (center). The complete model is formed by pinning the two limbs together at the hip (right).

The Froude number

The inverted pendulum on its own, an even simpler model than the ballistic walking model, does not accurately predict human walking speed but does provide some important insights about the physical constraints on how fast we can walk. Consider an inverted pendulum, as shown on the left side of Figure 2.9, of length ℓ where the body mass m is concentrated at the hip. When the hip is at its maximum height, its instantaneous velocity v is directed horizontally and the vertical reaction force (F) in the ankle joint is equal to the centripetal force:

$$F = \frac{mv^2}{\ell} \tag{2.5}$$

The centripetal force is the downward force that must be applied to the pendulum to prevent it from leaving the ground. Because the ground is unable to apply downward forces to the foot (unless you've stepped in glue), the centripetal force must be generated by the body weight, mg. Equating these forces and solving for velocity, we find that the maximum walking speed (v_{max}) for an inverted pendulum model is

$$v_{max} = \sqrt{g\ell} \tag{2.6}$$

We use v_{max} to define nondimensional walking speed (v^*):

$$v^* = \frac{v}{v_{max}} \tag{2.7}$$

At speeds above v_{max}, the foot will leave the ground; thus, $0 \leq v^* \leq 1$, to the extent that we walk as an inverted pendulum. The Froude number (Fr) is a dimensionless quantity that represents the ratio between the centripetal and gravitational forces:

$$\text{Fr} = \frac{v^2}{g\ell} = (v^*)^2 \tag{2.8}$$

The nondimensional walking speed and Froude number provide very useful metrics for comparing walking speeds between animals and among people with different maximum speeds and leg lengths.

An important prediction of the inverted pendulum model is that the maximum walking speed decreases if ℓ or g decreases. The first relationship is evident when watching a child walk alongside an adult. With shorter legs, the child will be walking at a higher Froude number than the adult and will have a lower v_{max}. Thus, you might observe an adult walking at a comfortable pace while a child is running to keep up. The dependence of v_{max} on gravity, g, explains the challenges that astronauts experienced on the lunar surface. Because the gravity on the moon is only about one-sixth that on earth, v_{max} on the moon is $\sqrt{1/6} \approx 0.4$ of its terrestrial value. Thus, a normal Earth-based walking speed will cause an astronaut to lift off the lunar surface. For this reason, Apollo 11 astronauts Neil Armstrong and Buzz Aldrin walked mostly at a slow pace to keep their feet on the ground. The later crews adopted a variety of bounding gaits. Interestingly, experiments before the moon missions had suggested that a kangaroo hop would be the most efficient gait, but the astronauts resorted to this strategy only rarely.

It is important to note that humans and other land animals do not normally walk at a Froude number of 1; they transition from walking to running at Froude numbers around 0.5 to save energy. An important exception is elephants, the largest land animal, which appear to walk much faster than the inverted pendulum model would allow, at Froude numbers greater than 1 (Hutchinson et al., 2003). As always, discrepancies between theory and observation provide an opportunity for learning. From 2003 to 2010, John Hutchinson and colleagues studied video and conducted experiments on elephants using custom-built force plates. (They had to be strong force plates!) Hutchinson discovered that even though the elephant always kept at least one foot on the ground and thus was still "walking" in the traditional sense, it went into an exaggerated crouch at high speeds (Figure 2.10). This "Groucho walk," named after comedian Groucho Marx who popularized this style of walking, meant that the leg muscles were generating high forces, and the exchange of gravitational potential and kinetic energy was inconsistent with the inverted pendulum model of walking. Thus, the elephants' gait was something of a blend between walking and running. We will explain this idea more fully in the next chapter, when we discuss the differences between walking and running and whether or not another large animal, the *Tyrannosaurus rex*, could run.

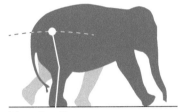
Typical walking

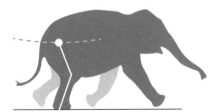
Fast Groucho walking

FIGURE 2.10
Even during a race, elephants always keep at least one foot on the ground but use a more bouncy gait (bottom) than during typical walking (top). This is often called "Groucho walking," made famous by Groucho Marx.

Cost of transport

The ballistic walking model captures some of the salient features of normal walking, in part because we naturally learn how to move in ways that minimize our energy expenditure. Reducing the activity of muscles during leg swing is one example. We also naturally select our walking speed, cadence, step width, and other variables to minimize cost of transport, or the energy required to move a given distance (Figure 2.11). Many studies over the past 60 years have verified that we walk in a way that minimizes our cost of transport—an important tenet of biomechanics. The cost of transport for walking is typically estimated by collecting and analyzing the mixture of gases inhaled and exhaled by a person while they walk at a particular speed (Figure 2.12).

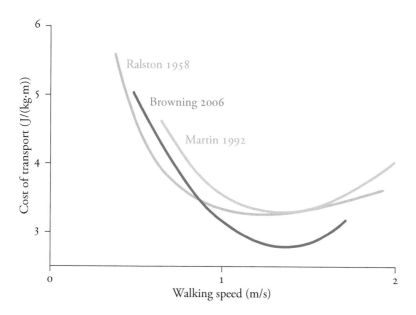

FIGURE 2.11
Cost of transport varies with walking speed. The energy required to move a unit mass a unit distance is lowest near the natural walking speed, as shown by three experimental studies. Data from Ralston (1958), Browning et al. (2006), and Martin et al. (1992).

Note that we cannot measure energy use directly, but indirect estimates based on measurements of breathing are an excellent proxy. Our muscles are powered by chemical reactions that burn carbohydrates, fats, and proteins. These reactions consume oxygen and produce carbon dioxide. By determining the rate at which exhaust is produced, we can infer how much fuel was used and

FIGURE 2.12
Experiment during which oxygen consumption and carbon dioxide production are measured to calculate the energy expended during running. The same type of experiment is used to calculate the cost of transport in walking shown in Figure 2.11.

therefore how much energy was consumed. As shown in Figure 2.11, we often express cost of transport as the number of Joules of energy required per kilogram of body mass to travel one meter.

Although the ballistic model can be useful for understanding some basic characteristics of walking, it has some limitations. For example, it is unable to model double support, which is essential for understanding the transition between consecutive steps. The inverted pendulum model predicts that our maximum walking speed will occur at a Froude number of 1, whereas in reality humans transition from walking to running at Froude numbers well below 1, and race walkers can "walk" at Froude numbers above 1. Also notice that a ballistic walking model might lead one to conclude that walking does not cost any energy. Furthermore, a model with a rigid stance limb does not accurately predict the ground reaction force observed experimentally. Finally, the ballistic walking model does not produce a repeated gait cycle. Some of these limitations are resolved in a slightly more complex model called the dynamic walking model.

Dynamic walking model

The ballistic walking model improved on the pendulum model by adding knees (or, to be more precise, one knee). We might expect, then, that adding one more anatomical ingredient would improve the model further. But it did more than that: by adding feet, Tad McGeer designed a walking robot that could be built and tested in a laboratory.

McGeer had been trained as an aeronautical engineer, so it is no surprise that he adopted a strategy echoing that used to develop the airplane a century earlier. Before they attempted a powered flight, the Wright brothers worked for years developing gliders that were powered by gravitational potential energy as they flew down a slope. By the end of 1902 they had completed hundreds of such flights. Once the Wrights had mastered gliding, they were confident they could master powered flight and recorded their first powered flights the following year (Collins et al., 2005).

Emulating the Wright brothers, ten years after Mochon and McMahon proposed the ballistic model, McGeer built a passive mechanism that could stably walk down a gentle slope, powered only by gravity, when provided with the right initial conditions (Figure 2.13). Proof that such a mechanism could be built was a breakthrough and started a new field of research into "dynamic walking," locomotion generated primarily by the passive dynamics of the legs. Today, powered dynamic walking robots are designed following the principle used by the Wright brothers: if a passive mechanism can move down a gentle slope strictly under the influence of gravity, then an active mechanism should be able to move on level ground using actuators that inject only the small amount of energy that would be supplied by gravity when descending a gentle slope.

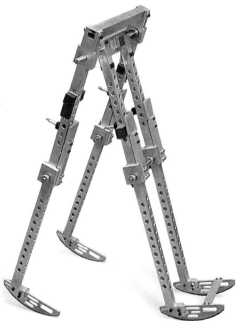

FIGURE 2.13
Replica of a walking machine described by Tad McGeer (1990). This mechanism, which includes two pairs of legs, will walk down a sloped surface when given the correct initial pose and push. Photo courtesy of Steve Collins.

The dynamic walking model extends the ballistic model in three key respects: by adding feet, modeling double support, and, most importantly, enabling the step-to-step transition (Figure 2.14). These features enable the mechanism to engage in cyclic, continuous walking. During single support, the foot of the stance limb rolls (without slipping) on the ground while the swing limb swings passively. Knee extension stops prevent hyperextension of the knee at the end of swing and keep the stance limb fully extended, passively supporting body weight. The center of mass moves upward in the first half of stance and downward in the second half. Just before foot contact, the center of mass is redirected to an upward trajectory by a push-off moment generated at the ankle of the stance limb. Redirecting the center of mass before foot contact reduces the velocity of the foot–ground collision and the corresponding energy loss. Some energy is still lost in the collision between the foot and

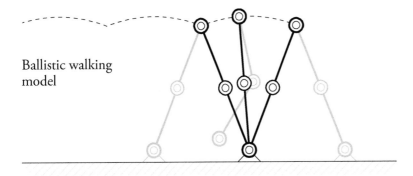

Ballistic walking model

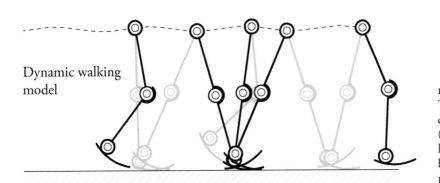

Dynamic walking model

FIGURE 2.14
The dynamic walking model described by Kuo and Donelan (2010). This model addresses some limitations of the ballistic model by enabling step-to-step transitions, producing a repeated gait cycle.

the ground, as well as when the knees hit the stops at full extension. To produce continuous gait, a small amount of energy must be injected in each step to compensate for this dissipation and for any frictional losses in the joints. In fully passive walking machines, this energy is provided by the gravitational potential energy as the mechanism walks down a gentle slope; on level ground, this energy can be injected by motors at the ankles or hips.

The dynamic walking model provides a theoretical framework for understanding some fundamental features of walking on level ground. Dynamic walking models have been used to study energy consumption over the gait cycle, how energy consumption changes with walking speed, and other aspects of human walking. For example, the energy required to redirect the center of mass trajectory at the beginning of double support is known as the step-to-step transition cost. The dynamic walking model correctly predicts that this cost increases with longer and faster steps. Increasing step length (while maintaining a constant step frequency) increases energy expenditure because the velocity of the center of mass increases and its trajectory must be redirected by a larger amount. Increasing step frequency (while maintaining a constant step length) increases energy expenditure because the legs must swing faster than would occur due to passive dynamics alone, and muscles expend energy to swing the legs.

The energy expenditure due to the step-to-step transition and forced swing leg motion contribute to the cost of human walking, along with maintaining balance. Humans usually select step lengths and step frequencies that minimize cost of transport. When increasing walking speed beyond the speed at which cost of transport is lowest, humans increase step length and step frequency in nearly equal proportion, balancing increases in cost due to the step-to-step transition and forced leg motion. The step-to-step transition cost is also mitigated by using our feet like sections of wheels (note the arc-shaped feet in the lower part of Figure 2.14), reducing the required change in the direction of the center of mass. The human foot effectively rolls over the ground as the ground reaction forces move from the rear of the foot at foot contact to the forefoot at toe off (Figure 2.5). In a dynamic walker, the position of the foot is essential: if the foot is placed too far forward, the walker will fall backward; if the leg swings too slowly and the step length is too short, the walker will fall forward.

The dynamic walking model has more analytical strength than the ballistic model, but of course still has limitations. Many modeling assumptions and simplifications preclude using the model to study certain elements of human walking. For example, the dynamic walking model assumes strictly planar motion, neglecting motion out of the sagittal plane. McGeer's apparatus enforces this assumption by doubling the number of legs to minimize side-to-side sway. In 2001, Steve Collins, then an undergraduate at Cornell University, built the first two-legged passive dynamic walking machine (Figure 2.15). Designing this mechanism required careful attention to the trajectory of the machine's center of mass, to prevent it from falling due to non-sagittal-plane motion. Counter-swinging arms in the sagittal plane stabilized yaw, foot shape and lateral arm swing controlled lean, and a soft heel avoided sensitivity to the pose at foot contact (Collins et al., 2001). In each case, biomimicry improved the performance of the mechanism, and it gives us some idea of why each of these ingredients—arm swing, feet, and heels—give humans a biomechanical advantage.

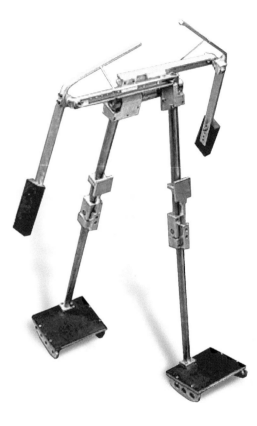

FIGURE 2.15
The first two-legged passive dynamic walker. Photo courtesy of Steve Collins.

Arm swing

Thus far, we have used simple models to study the dynamics of the lower body, but the upper body plays a role in walking as well. The most obvious motion in the upper body is arm swing, but why we swing our arms when we walk is less obvious. To study this behavior, Steve Collins and his colleagues proposed a straight-legged passive walking mechanism similar to McGeer's but with arm-like pendula attached at the hips (Figure 2.16). They tested several arm swing strategies, including the familiar "normal" swing, and an "anti-normal" swing where the left arm advances with the left leg and the right arm with the right leg. I can't help thinking of this

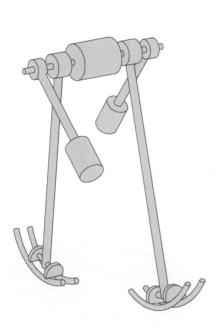
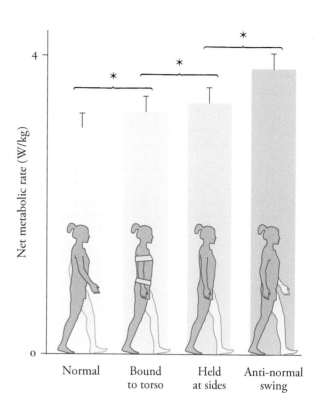

strategy as the "Bro walk," in honor of my older brother Brian Delp, who enjoyed walking this way for his and our amusement when we were children.

Experiments on human subjects walking normally showed that the activity of muscles crossing the shoulder and elbow joints is low, confirming that normal arm swing is primarily passive. The kinematics and kinetics of the lower body are unaffected by arm swing strategy, but the whole body twists more about the vertical axis when the arms are not swinging, and still more during anti-normal arm swinging. The increase in angular momentum is countered by a higher ground reaction moment about the vertical axis, with a corresponding increase in muscle activity and energy expenditure. To put it more simply, the Bro walk is awkward, and that's why people past a certain age don't usually walk this way.

FIGURE 2.16
Normal and anti-normal arm swing emerge spontaneously in a passive walking mechanism with arms (left). In human experiments, energy expenditure was least using normal arm swing and highest using anti-normal arm swing (right; error bars and asterisks indicate standard deviations and statistically significant differences). Adapted from Collins et al. (2009).

Skeletal model for gait analysis

Because each biologically inspired feature of passive walkers has improved their performance, it's natural to wonder what a higher-fidelity model would look like. Let's start with some bad news: anatomical joints are complex, with adjacent body segments translating and rotating relative to each other in all directions. We are often interested in, or can only accurately measure, a subset of these motions. Therefore, even a model deliberately patterned after the human body will make simplifying assumptions, such as disallowing relative translation between the femoral head and the pelvis so that the hip is represented as a ball-and-socket joint.

Nevertheless, a typical and reasonably accurate lower-limb skeletal model used for analyzing gait is shown in Figure 2.17. (It may be supplemented with a torso and arms when necessary.) The model consists of 9 articulating rigid bodies: a pelvis and left and right femur, patella, tibia-plus-fibula (shank), and foot. The model has 16 degrees of freedom in the lower body: 6 describe the position and orientation (tilt, list, and rotation) of the pelvis relative to a fixed frame of reference, and 5 describe the pose of each leg (hip flexion, adduction, and rotation; knee extension; and ankle dorsiflexion). These terms are standard, well worth learning, and shown in Figure 1.16.

Just think about this for a moment: at every instant while we are walking, we are simultaneously controlling more than a dozen angular positions, without being aware of it. By contrast, the ballistic walking model of Mochon and McMahon had only 3 degrees of freedom, and the dynamic walking model had only a few more. We succeed in part because we take shortcuts, using passive motions whenever possible, but also because we have a brain that spends a great deal of time in the first year of our life learning how to coordinate the complex movement of walking. Next time you joke that someone can't walk and chew gum at the same time, remember that even the first part—bipedal walking—is an amazing accomplishment.

Notice that the model shown in Figure 2.17 is three-dimensional. As we have seen, walking is a three-dimensional activity, and

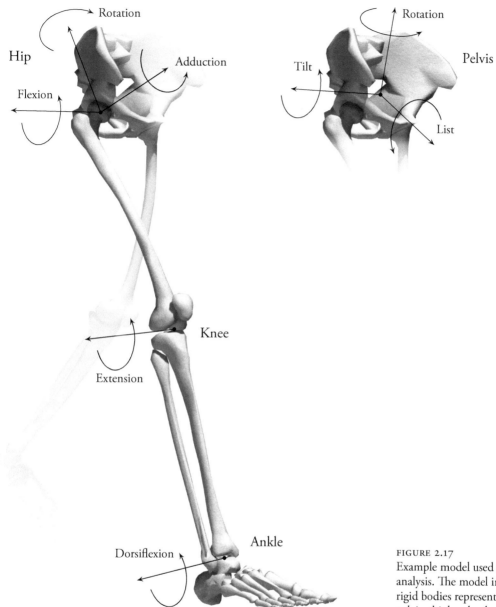

non-sagittal-plane motions and forces must be analyzed to understand balance and body-weight support. A variety of methods are available to measure these three-dimensional motions, which we describe in Chapter 7. We can use these methods to estimate the motions of human subjects during walking, which we present next.

FIGURE 2.17
Example model used in gait analysis. The model includes rigid bodies representing the pelvis, thighs, shanks, and feet, and permits only those joint motions that can be reliably measured in a typical gait analysis: pelvic tilt, list, and rotation; hip flexion, adduction, and rotation; knee extension; and ankle dorsiflexion. Adapted from Rajagopal et al. (2016).

Kinematics of walking

The pelvis undergoes a complex motion during walking (Figure 2.18). In the frontal plane, as the limb is loaded in early stance, the pelvis tips down on the swing side. The range of this motion nearly doubles from about 5 degrees at low walking speeds to about 10 degrees at high speeds. In the transverse plane, the pelvis rotates toward the advancing limb; thus, the hip of the lead limb is in front of the hip of the trailing limb. This motion also increases with speed and provides a mechanism for increasing step length.

The joints of the lower limb also move in stereotypical patterns (Figure 2.19). At foot contact, the hip is in flexion. During the

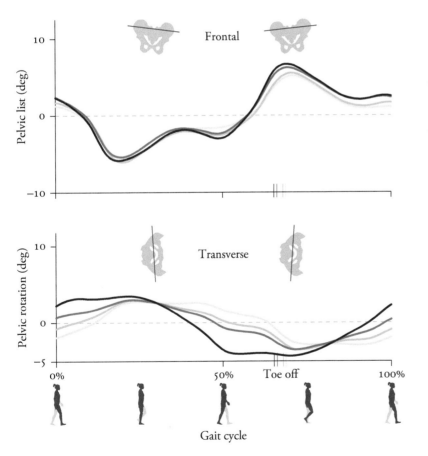

FIGURE 2.18
Representative pelvis orientations over the gait cycle when walking at several speeds. Averaged over 10 subjects. Vertical lines on the horizontal axis indicate toe off at each speed. Data from Arnold et al. (2013).

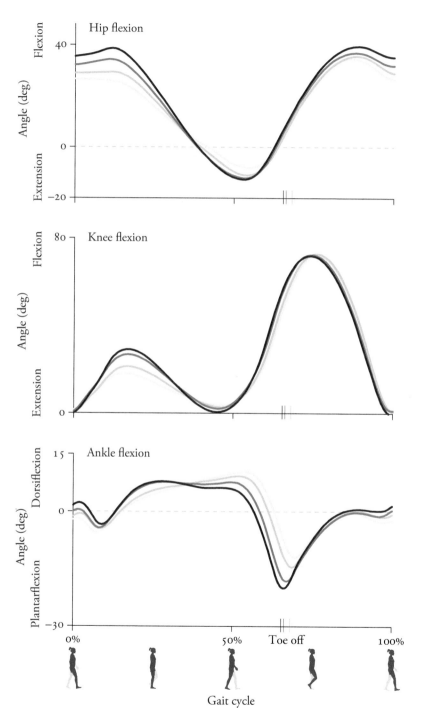

FIGURE 2.19
Joint motions over the gait cycle when walking at several speeds. Averaged over 10 subjects. Vertical lines on the horizontal axis indicate toe off at each speed. Data from Arnold et al. (2013).

stance phase the hip extends, reaching peak extension just prior to toe off, and then flexes during the swing phase. The knee is fully extended at foot contact. As the limb is loaded, the knee flexes and then extends, like a shock absorber, with larger flexion at higher speeds. Just prior to swing, the knee flexes rapidly, reaching peak flexion near mid-swing, and then extends rapidly to reach full extension before the next foot contact. The ankle is in a neutral position at foot contact. The ankle plantarflexes as the foot rotates toward the ground and then dorsiflexes as the tibia passes over the foot during early stance. Near the end of the stance phase, the ankle plantarflexes rapidly, reaching a peak at approximately the time of toe off. As walking speed increases, peak joint angles generally increase, and the fraction of the gait cycle spent in stance decreases.

Ground reaction forces and walking speed

The ground reaction forces during walking are shown in Figure 2.20. At high speed, the vertical component of the ground reaction force rises rapidly after foot contact and has a characteristic double-humped shape. We saw a similar shape in Figures 2.4–2.6. The humps become more pronounced as speed increases. The first hump of the ground reaction force arises from the muscles of the leading limb that support body weight. The second hump of the ground reaction force arises from the muscles of the trailing limb during push off. These reaction forces provide insight into the acceleration of the center of mass, but experiments alone cannot tell us which muscles are responsible for generating the measured ground reaction forces. We will examine how muscles coordinate walking and generate the ground reaction forces in Chapter 11.

The data shown in Figures 2.18–2.20 are freely available for download from simtk.org. Normative data such as these are valuable for quantifying aberrations in a subject's gait and for testing the accuracy of models and simulations of movement.

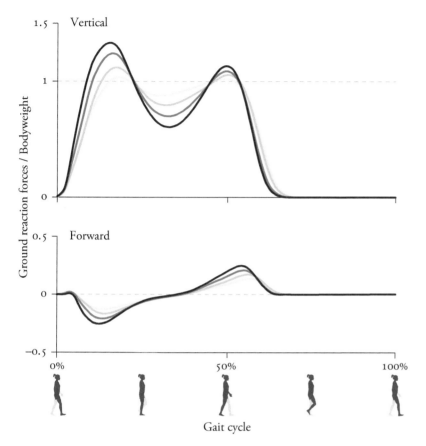

FIGURE 2.20
Representative ground reaction forces over the gait cycle when walking at several speeds. Normalized by body weight and averaged over 10 subjects. Data from Arnold et al. (2013).

Atypical gait

Humans can walk smoothly and efficiently using the mechanisms discussed in this chapter. However, injuries to the body or brain and diseases of the joints or muscles can disrupt walking dynamics. For example, individuals with cerebral palsy, a movement disorder that arises from damage to the brain, frequently walk with a crouch gait (Figure 2.21). Crouch gait is characterized by excessive flexion of the knee during the stance phase. Excessive knee flexion is problematic because it increases forces in the knees during stance, impedes toe–ground clearance during swing, and dramatically increases energy expenditure. (Try walking in a crouch gait for two minutes to see whether you get out of breath.) Excessive knee flexion typically worsens over time in individuals with cerebral palsy, often leading to altered knee joint mechanics and chronic knee pain. In severe cases, knee flexion may become so great that the person loses the ability to walk entirely.

Many persons with cerebral palsy, as well as individuals who have had a stroke, walk with a stiff-knee gait in which swing-phase knee flexion is diminished and delayed (Figure 2.22). This gait pattern also hinders toe–ground clearance and can result in tripping or require compensatory movements that are energetically inefficient. Stiff-knee gait is thought to be caused primarily by inappropriate

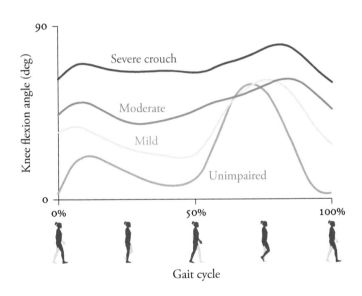

FIGURE 2.21
Crouch gait is characterized by excessive knee flexion during stance. Average knee flexion angle is plotted over one gait cycle for subjects who walked with an unimpaired gait and for subjects with cerebral palsy who walked with mild, moderate, and severe crouch gait. At left, Rachel Jackson demonstrates walking with a moderate level of crouch gait. Plot adapted from Steele et al. (2012).

activity of the rectus femoris muscle, which crosses in front of the knee via the patella and produces a knee extension moment. Accordingly, stiff-knee gait is often treated with a rectus femoris transfer surgery in which the attachment of this muscle is relocated from the patella to a site that decreases the muscle's ability to generate a knee extension moment. Unfortunately, outcomes of the rectus femoris transfer surgery are inconsistent: some individuals

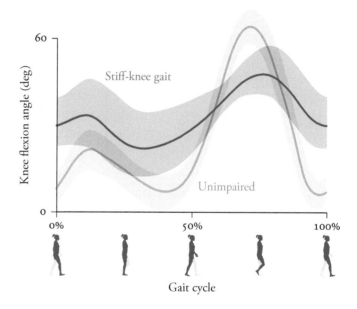

FIGURE 2.22
Stiff-knee gait is characterized by diminished and delayed knee flexion during swing. Knee flexion angle over the gait cycle is shown for individuals walking with stiff-knee gait compared to unimpaired gait (mean ±1 standard deviation). At left, Scott Uhlrich demonstrates walking with stiff-knee gait. Plot adapted from Fox et al. (2009).

show substantial improvement in their swing-phase knee flexion after surgery, but others show little change.

It is not possible to understand the causes of crouch gait or stiff-knee gait simply by examining joint kinematics, because measurements of a motion do not establish what caused the motion. As we will see in Chapter 11, muscle-driven simulations of walking can provide insights for understanding the causes of crouch gait and stiff-knee gait, and can be used to design effective treatments.

Changes in walking under various conditions

Walking is a stereotypical activity, which means that we can learn a great deal about it from simple models and average experimental data. Nevertheless, there is no single ideal gait. We all have had the experience of recognizing a friend from a distance just by the way he or she walked. In addition, there are many variations of walking that are completely "normal" adaptations to different circumstances. People walk differently when going slower or faster, carrying groceries, trekking uphill, marching in a parade, or wearing high-heeled shoes. We can also observe changes in walking dynamics when wearing robotic systems that have been optimized to decrease the energy we consume during walking. Table 2.1 lists some of the changes that have been observed when walking in various conditions, all of which increase the cost of transport, with the exception of walking downhill. There are, of course, many more changes that have been observed during walking in steady-state conditions as well as transient conditions like gait initiation, turning, and accelerating.

Humans are efficient walkers. In the best case, we simply fall forward and then step to catch ourselves from falling, injecting a little energy to make the transition between steps and to swing our legs. Yet there are many cases in which walking is impaired, which can limit activities of daily living. Walking is the primary form of physical activity for billions of individuals, and there are severe health consequences when physical activity is limited. To help restore and improve walking in the face of neurological and musculoskeletal diseases, it is essential to have a deep understanding of walking dynamics. This chapter has just scratched the surface. Later, we will examine the actions of muscles in normal and impaired walking. For now, we transition from walking to running.

TABLE 2.1
Changes observed when walking in various conditions

Condition	Observed changes
Slow	Decreased knee flexion during stance and swing, modulation of ground reaction force, and muscle activity
Fast	Increased knee flexion during stance and swing, modulation of ground reaction force, and muscle activity
Crouch gait	Increased knee flexion during stance, commonly accompanied by increased hip flexion
Stiff-knee gait	Decreased and delayed peak knee flexion in swing
Drop foot	Decreased ankle dorsiflexion in swing
Equinus gait	Increased ankle plantarflexion in stance (walking on the toes)
High-heeled shoes	Increased ankle plantarflexion, knee flexion, and loading rate on the supporting limb
Uphill	Increased hip flexion, knee flexion, and ankle dorsiflexion during stance
Downhill	Increased ankle plantarflexion during early stance; increased hip and knee flexion during late stance
Slippery surface	Decreased angular velocity of the foot at initial contact, loading rate on the supporting limb, and stride length
Carrying a backpack	Increased stance duration, peak flexion angles during stance, ground reaction force, and muscle activity
Added leg mass	Increased cost of transport, with larger increases for more distal mass location; increased activity in muscles that initiate swing
Obesity	Increased step width, ground reaction force, and sagittal-plane net muscle moments during stance
Below-knee prosthesis	Increased hip extensor activity during early and midstance

3 Running

If you can fill the unforgiving minute
With sixty seconds' worth of distance run . . .
—Rudyard Kipling

ANYONE WHO HAS WATCHED *Jurassic Park* probably remembers the iconic scene where a Jeep tries to outrace a pursuing *Tyrannosaurus rex*. Interestingly, the scene agreed with the conventional wisdom of paleontologists at the time of the movie (1993). It was believed that *T. rex* could run at speeds of 40 km/h, and some scientists argued it could run even faster. At such speeds it could plausibly have outpaced a Jeep on an unpaved road.

However, more recent biomechanical studies, beginning with John Hutchinson's groundbreaking article in 2002, have shown that *T. rex* probably couldn't run at all. Even if it could, it likely could not have reached 40 km/h. Using simulations of *T. rex*, Hutchinson showed that to achieve such a speed, 86 percent of its body mass would have had to be leg muscle, leaving little room in the budget for its massive tail, head, and torso. It would have required dedicating this much mass to leg muscles in order to generate sufficiently large ground reaction forces, which typically exceed 2 bodyweights when running.

Studies of animals like dinosaurs, elephants, and kangaroos help to clarify our understanding of human running: what drives the transition from walking to running, and what constrains running speed. Measurements of ground reaction forces show why more injuries occur during running than walking, and how we might design running tracks to decrease injury rates and increase speed.

In this chapter, we will explore these questions using simple mechanical models. The models include springs to represent the elastic properties of muscles and tendons, and reveal how the storage and release of elastic energy makes running efficient. To get started, we first define the running gait cycle and study the forces and elastic mechanisms involved. We then discover principles that will help you design tracks, shoes, and prosthetic limbs that enable fast and efficient running. We also examine the changes in gait and energy expenditure that occur as we transition from walking to running.

The running gait cycle

The running gait cycle comprises alternating phases of single-leg support and flight (Figure 3.1). As with walking, a running gait cycle is defined by two consecutive foot contact events on the same leg, with foot contact of the opposite leg occurring halfway through. Each leg has a stance phase, when the foot is in contact with the ground, and a swing phase, when the foot is off the ground. Stance begins at foot contact, ends at toe off, and accounts for about 30 to 45 percent of the gait cycle in human running, though it may account for 25 percent or less during high-speed sprinting. Recall that the duration of the stance phase in walking decreases as walking speed increases. Thus, as walking speed increases, the duration of double-support shortens. Increasing speed even further, the stance

FIGURE 3.1
The running gait cycle and its constituent events (e.g., foot contact) and phases (e.g., support). The percentages of time spent in stance and swing vary with running speed and style.

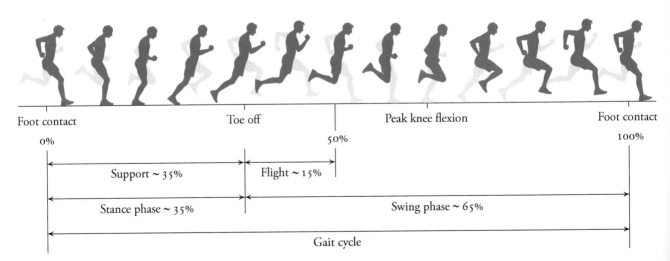

phase corresponding to each leg will eventually span less than half the gait cycle, giving rise to a flight phase.

The presence of a flight phase is one way we distinguish walking from running. In fact, it is quite tempting to believe that it is the signature of running, not only in humans but in other animals. However, we will see shortly that another qualitative change occurs when we switch from a walking gait to a running gait, which is even more important from a biomechanical point of view.

The metrics used to quantify the running gait cycle are similar to those used for walking. The step length is the distance along the line of progression between two consecutive footprints. The distance traveled in two consecutive steps, or the distance covered in one gait cycle, is called the stride length. The rate at which foot contact events occur (equivalently, the reciprocal of step duration) is called the step frequency or cadence; the rate at which strides are taken is called the stride frequency. Running speed can be computed as the product of stride length and stride frequency or, equivalently, as the product of step length and cadence:

$$\text{speed [m/s]} = \text{step length [m/step]} \times \text{cadence [steps/s]} \quad (3.1)$$

A moderate running speed is 4 m/s. At this speed, the stance phase accounts for about 35 to 40 percent of the gait cycle, a typical cadence is 180 steps/min, and a typical step length is around 1.3–1.4 m:

$$\text{step length} = \frac{4.0 \text{ m}}{1 \text{ s}} \times \frac{60 \text{ s}}{1 \text{ min}} \times \frac{1 \text{ min}}{180 \text{ steps}} = 1.33 \text{ m/step} \quad (3.2)$$

Of course, these quantities will vary with leg length, running style, footwear, and other factors.

Ground reaction forces

As we saw in Chapter 2, we can gain insight into the dynamics and energetics of gait using measurements of the ground reaction forces over time. During running, the vertical component of the ground reaction force rises rapidly following foot contact and reaches a maximum at about 15 to 20 percent of the gait cycle (Figure 3.2).

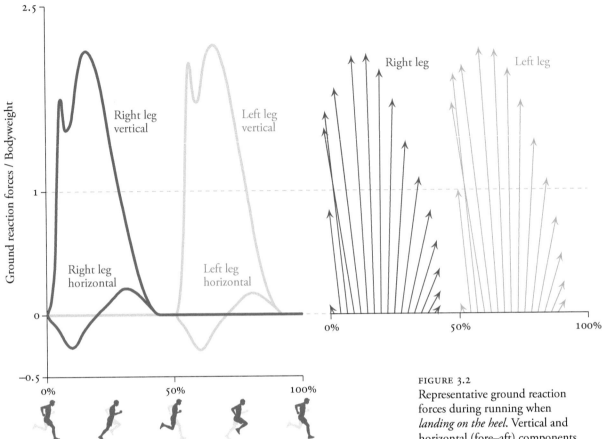

FIGURE 3.2
Representative ground reaction forces during running when *landing on the heel*. Vertical and horizontal (fore–aft) components (left) and schematic of total vector (right) are shown over the gait cycle. *A sharp peak* in the vertical ground reaction force occurs at heel strike. Data from Yong et al. (2020).

The vertical ground reaction force peaks at around 2 bodyweights when running at a moderate speed. The horizontal ground reaction force points backward during the first half of stance, decelerating the center of mass, and points forward thereafter.

Figure 3.2 illustrates the characteristic double-peaked vertical ground reaction force observed in rearfoot strikers, runners who land on their heels. Although the ground reaction force in walking also has two peaks, in running the first peak is shorter and is due to the impact of the heel on the ground. Certain runners, notably those who have grown up running barefoot, land instead on their forefeet. For these runners, the vertical ground reaction force

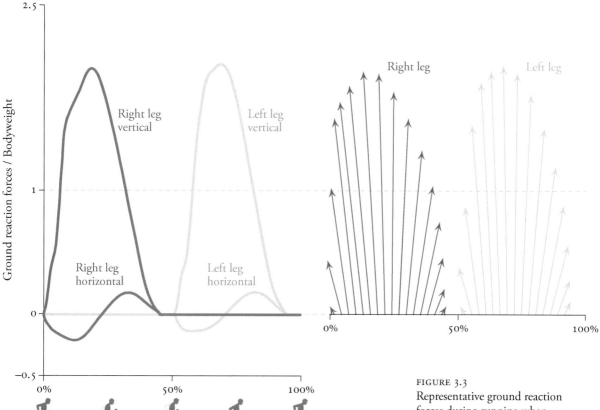

FIGURE 3.3
Representative ground reaction forces during running when *landing on the forefoot*, from the same subject as in Figure 3.2. The vertical ground reaction force is devoid of the sharp peak observed when rearfoot striking, which may reduce injury risk. Data from Yong et al. (2020).

rises more gradually and has only one peak (Figure 3.3). It has been argued that this style of running is more "natural" and protects against impact-related injuries, but subsequent research has cast doubt on this conclusion. A runner who has grown up rearfoot striking may risk injury if he or she changes to forefoot striking, especially without proper training.

The forward kinetic energy and gravitational potential energy during running can be computed using Equations 2.3 and 2.4 (Figure 3.4). The forward speed and forward kinetic energy are the greatest, and approximately constant, during the flight phase of running (neglecting air resistance). The center of mass, and therefore

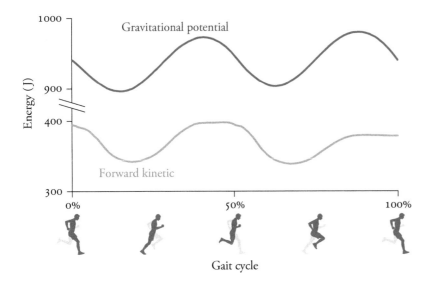

FIGURE 3.4
Representative gravitational potential and forward kinetic energies during running. Forward kinetic energy is constant during flight. Data from Yong et al. (2014).

the gravitational potential energy, is also highest at midflight. Thus, unlike in walking, the forward kinetic energy and gravitational potential energy reach their maxima at approximately the same time. Similarly, both forward kinetic energy and gravitational potential energy reach minima in midstance—that is, the forward speed is least approximately when the center of mass is lowest.

We want to emphasize a fundamental difference between walking and running: in running, we do not benefit from the exchange of forward kinetic energy and gravitational potential energy, but instead store and release elastic potential energy in our muscles and tendons as they stretch and recoil. This observation suggests a model of running as shown in Figure 3.5. In this model, the mass reaches its lowest point at midstance when the forward velocity of the mass center is also a minimum, similar to what we find in the experimental data. The spring-like action of our legs makes running an energetically efficient gait.

As we saw in Chapter 2, elephants do not have a flight phase, even when they are moving as fast as possible. For this reason, you might argue that elephants don't run. But in fact, they seem to have a "Groucho walk" in which their knees bend and fluctuations in their gravitational potential energy and forward kinetic energy become synchronized. From a biomechanical perspective, this synchronization means that the elephants are running in a sense. Just like people, the pachyderms are using their legs as springs to store energy, as illustrated in Figure 3.5. Most likely, elephants evolved this method of "running" without going airborne as an adaptation to their great size. As we will see later in this book, muscle strength scales with the cross-sectional area of muscle (a length squared), whereas body mass scales with volume (a length cubed). Thus, very large animals have a lower strength-to-weight ratio compared to smaller animals. This principle explains why small animals like squirrels can generate large ground reaction forces and leap several body lengths, but elephants and *Tyrannosaurus rex* (which are roughly the same size) cannot generate a ground reaction force sufficient for even an instant of flight.

FIGURE 3.5
The stance phase of running (left) and a mass–spring model thereof (right). In this model, the mass of the body is lumped into a point mass, which sits atop a massless linear spring representing the leg. The spring compresses and the mass reaches its lowest point during midstance, when the forward velocity of the mass is also lowest.

Elastic mechanisms in hopping and running

Another animal that can teach us about elastic energy storage during running is the kangaroo. Kangaroos do not run in the conventional sense of the word; normally we describe their motion as a hop. Nevertheless, from a biomechanical point of view, they use a similar mechanism to the one humans use when running.

In the early 1970s, Terence Dawson and Richard Taylor trained kangaroos to wear a facemask while hopping on a treadmill at

Pentapedal gait

Hopping

High-speed hopping

speeds ranging from 1 to 22 km/h—an experiment that surely required virtuosic animal-handling skills. The facemask allowed the researchers to measure the animals' metabolic energy consumption, which was the main theme of Taylor's research at the time. His many experiments with kangaroos and other animals focused biologists' attention on energy consumption as a key factor in animal behavior.

At low speeds, Dawson and Taylor found that kangaroos do not hop but instead move with a pentapedal gait in which they use their tail as a fifth point of contact with the ground (Figure 3.6). This manner of locomotion looks awkward and is energetically inefficient. Indeed, they found that energy consumption increased dramatically as the animals hobbled forward at higher speeds (from 1 to 6 km/h; Figure 3.7A). The increase in speed during pentapedal gait was achieved primarily by increasing stride frequency (Figure 3.7D). At about 6–7 km/h, kangaroos transition to hopping. A feature that jumps out of Figure 3.7A is the slight *decline* in the cost of transport as the animals' speed increases above 7 km/h. As Figures 3.7C and D show, in the hopping phase they increase their speed primarily by increasing stride length. Stride frequency is nearly constant. This result is consistent with a mass–spring model of the kangaroo (similar to Figure 3.5), because the natural frequency of a sprung mass does not change when the amplitude of its motion

FIGURE 3.6
Modes of locomotion in kangaroos. During slow locomotion, kangaroos adopt a pentapedal gait (top) in which they use their tail for support when advancing their hind limbs. At higher speeds, kangaroos adopt a hopping gait (middle) in which their tail is used for balance and to control body pitch. During high-speed hopping (bottom), kangaroos can reach speeds in excess of 50 km/h. Adapted from Dawson (1977).

changes. Dawson and Taylor noted that the Achilles tendon of the kangaroo is well suited to store and release elastic energy, and suggested that this mechanism makes hopping efficient and enables hopping at high speed.

Around this time, Taylor learned about the force plates in Giovanni Cavagna's laboratory in Italy, and he brought his menagerie (kangaroos, monkeys, dogs, and turkeys) to Milan to try them out on Cavagna's then-new invention. The resulting study, published in 1977, provided additional evidence that kangaroos

FIGURE 3.7
Energetics of kangaroo locomotion. Mass-normalized rate of oxygen consumption increases with increasing speed during pentapedal gait. Kangaroos transition to hopping at around 6–7 km/h; oxygen consumption then decreases with increasing speed until a minimum is reached at about 20 km/h. Dawson (1977) estimated oxygen consumption for speeds that could not be studied in the lab (dashed lines). At low hopping speeds, stride frequency remains relatively constant and speed is increased primarily by increasing stride length. Data from Dawson (1977).

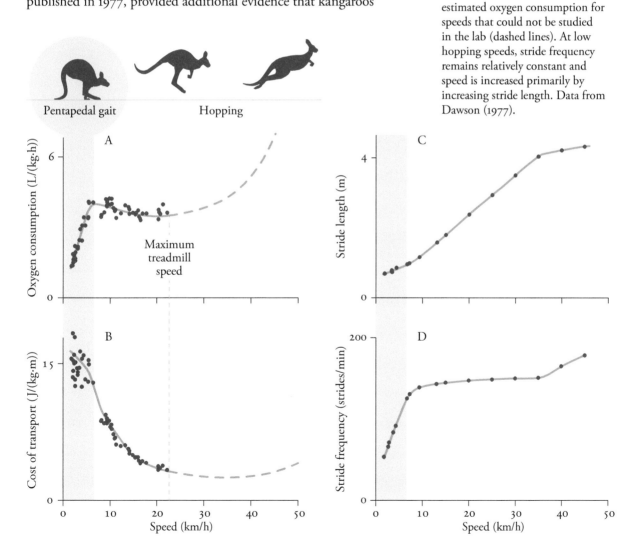

store elastic energy and release it during hopping. At speeds of up to 30 km/h, oxygen consumption accounted for about one-third of the energy required to accelerate the kangaroo's center of mass; the remaining energy must have been supplied by elastic storage in muscles and tendons.

Hopping robots

Many early walking robots remained statically stable during locomotion, shuffling forward while keeping their center of mass above their feet. In the 1980s, Marc Raibert and his colleagues built the first robots that maintained "dynamic balance" during high-speed motion, moving with alternating phases of support and flight as in running. These early robotic runners had only one leg and relied on motion to maintain stability, remaining upright only by hopping. This clever design focused the control problem on maintaining balance while avoiding the complexity of coordinating the motions of multiple legs.

The first of Raibert's robots hopped in a plane on a single springy leg (Figure 3.8). The robot was mechanically constrained to prevent it from falling on its side and was tethered to a control computer and offboard supplies of compressed air and electrical power. The

FIGURE 3.8
The motion of Raibert's planar robot resembles that of a hopping kangaroo, albeit without a tail to help control pitch. Onboard electronics adjust hip angle and leg stiffness to control balance, vertical motion, and body attitude. Unlike the marsupial, the machine can balance only while it hops. Adapted from Raibert and Sutherland (1983).

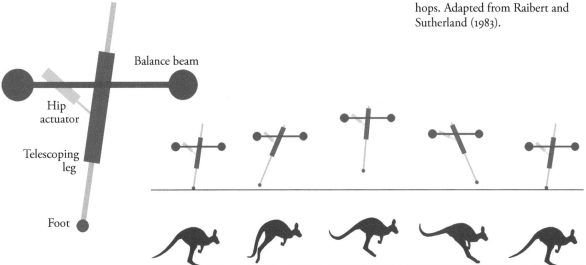

leg bounced on a spring (pneumatic cylinder) with adjustable stiffness, which provided the energy required for each hop. The controller comprised three independent components: the first regulated vertical motion by adding air to and releasing air from the pneumatic cylinder, adjusting leg stiffness; the second maintained balance by appropriately positioning the foot during flight; and the third stabilized the body's attitude (inclination) by applying torque between the leg and body during stance. Leg swing was a natural consequence of the hip torques generated by the balance and attitude controllers. For example, during flight, the balance controller must swing the leg forward to prepare for the next stance phase.

The same three-part control strategy was used to dynamically balance an analogous 3D hopping robot (Raibert et al., 1984). The robot again bounced on a single springy leg, storing and releasing energy with each hop. Unlike its planar predecessor, however, the body was unconstrained and thus the balance controller regulated both its pitch and roll angles. The robot reached a top speed of 2.2 m/s (a slow running pace) and was able to recover from external perturbations, proudly supplied by the robot's designers. Raibert went on to found Boston Dynamics in 1992, which has developed a series of agile robots. They have graduated to two and four legs, but they continue to be based on the same principles. After nearly two decades of developing cutting-edge robots, the company announced its plans to make its first commercial model, the canine-inspired SpotMini, in 2019 (Figure 3.9).

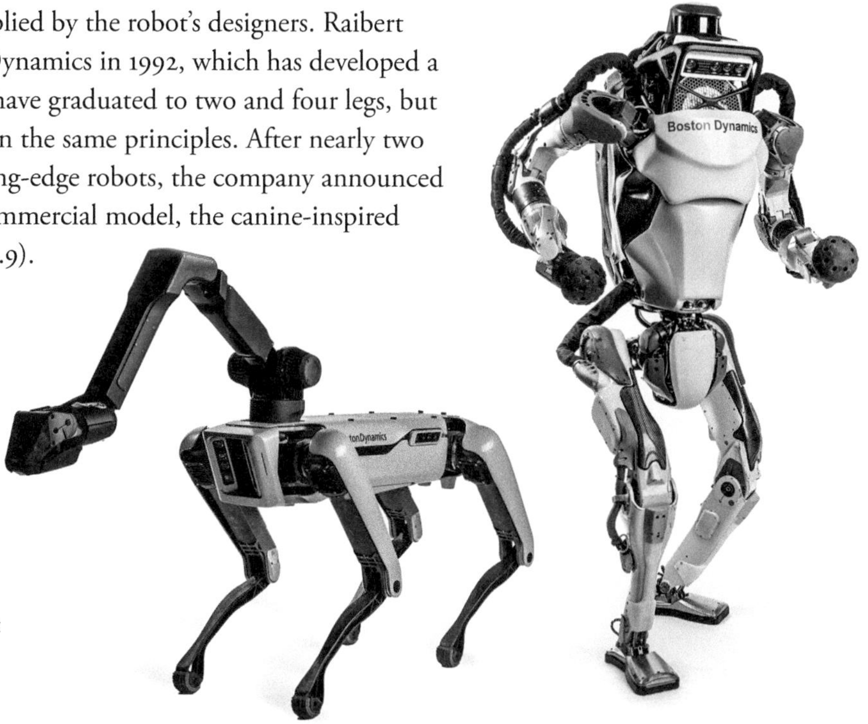

FIGURE 3.9
Examples of modern running robots. Photos of the Spot robot (left) and Atlas robot (right) are courtesy of Boston Dynamics.

Tuned track

In the mid-1970s, Thomas McMahon received an unusual call for help. Harvard University was building a new indoor track, and the contractor disagreed with the track coach on how steeply banked the turns should be (Wingerson, 1983). After McMahon resolved that issue (in favor of the track coach), they asked him another question: How hard should the track surface be?

At that time, the conventional wisdom was the harder the better, to minimize the amount of time the runners' feet would be in contact with the ground. But McMahon questioned this belief. It seemed to him that a more compliant (or "springy") surface might increase runners' speed by increasing stride length, stride frequency, or both (recall Equation 3.1). He also theorized that a compliant track might reduce the risk of injury by decreasing the initial spike in the ground reaction force observed in rearfoot-striking runners (Figure 3.2).

A mathematical model can shed light on this theory. Together with his colleague Peter Greene, McMahon focused on the stance phase of running, the only part of the gait cycle when the track can directly influence the runner's dynamics. They modeled the runner as a mass–spring–damper system in series with a spring representing

FIGURE 3.10
Conceptual model used by McMahon and Greene for predicting running performance on a compliant track. A passive spring–damper system represents the mechanical properties of the muscles, and a spring under the foot represents the compliance of the track. The dynamics of this system as it compresses during early stance (center) and rebounds during late stance (right) are analyzed to find the stiffness of the track spring that is best tuned to the properties of the leg. Adapted from McMahon and Greene (1979).

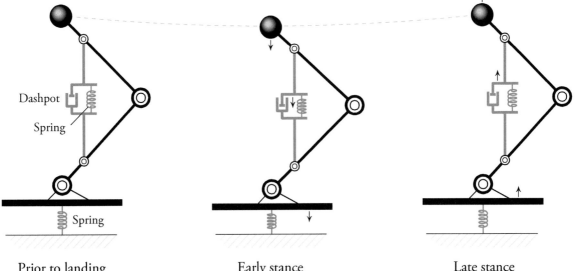

Prior to landing   Early stance   Late stance

the track (Figure 3.10). The runner's spring reflects the elastic properties of muscles and tendons; the damper reflects the influence of velocity on muscle force. The final component of the model is a linear spring that represents the compliance of the track and whose optimal stiffness was sought.

This system has two degrees of freedom and its dynamics are governed by two second-order ordinary differential equations, which can be derived from the schematic shown in Figure 3.11:

$$m\ddot{x}_r + b_m(\dot{x}_r - \dot{x}_t) + k_m(x_r - x_t) = 0$$
$$b_m(\dot{x}_r - \dot{x}_t) + k_m(x_r - x_t) - k_t x_t = 0 \quad (3.3)$$

where m is the mass of the runner's body, x_r is its displacement from equilibrium, x_t is the displacement of the track under the foot, b_m is the resistance of the damper, and k_m and k_t are the stiffnesses of the muscle and track springs, respectively.

McMahon and Greene estimated the foot–ground contact time as a function of track stiffness k_t by computing the natural frequency of this system and noting that the foot will lift off the ground after approximately half the period of one oscillation. Their first unexpected finding was a range of intermediate track stiffnesses for which the contact time is less than that when running on a very stiff surface, such as concrete (t_0; Figure 3.12). Even if the stride length stayed the same, this finding would lead one to expect an increase in running speed because stride frequency tends to increase as the contact time decreases.

But there was even better news to come. Their model was too simple to predict stride length, so McMahon and Greene built running surfaces of various stiffnesses, ranging from nearly rigid to very compliant trampoline tracks, and estimated step lengths as individuals ran on these surfaces. As expected, running on very compliant surfaces dramatically increased step lengths compared to running on stiff surfaces (Figure 3.13). But of course, running on a trampoline is much too slow because of the greatly increased time of contact between the foot and the trampoline. The most relevant finding was that step lengths still increased by about 1 percent even in the "tuned region" (shown with a gray stripe in Figures 3.12 and

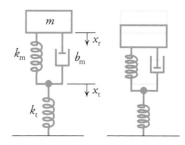

FIGURE 3.11
Schematic used to derive equations of motion for the conceptual model shown in Figure 3.10. The system is shown in equilibrium (left) and during stance (right). The runner's body is modeled as a mass (m) that moves with displacement x_r due to forces generated by a spring (stiffness k_m) and dashpot (damping b_m) that represent the combined effect of the leg muscles. The small piece of the track under the foot is modeled as a massless body that is connected to the ground with a spring (stiffness k_t) representing the track compliance. Adapted from McMahon and Greene (1979).

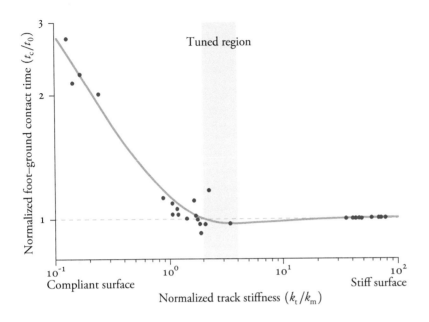

FIGURE 3.12
Foot contact time vs. track stiffness. Both axes are logarithmic scale. Foot contact time (t_c) is normalized by the contact time on a hard surface (t_0); track stiffness (k_t) is normalized by leg stiffness (k_m). The solid line was determined from the model shown in Figure 3.11 using a damping ratio of $b_m/(2\sqrt{mk_m}) = 0.55$ because this value gives the best fit to the experimental data. The tuned region (shaded) shows the potential to reduce contact time. Adapted from McMahon (1984).

3.13), where contact time also decreased. This is a win-win zone where both stride length and stride frequency are enhanced. For a track whose stiffness is within this region, roughly 2 to 4 times the stiffness of the runner's leg, we expect running speed to increase. (For comparison, concrete is 100 times as stiff as a human leg.)

Indoor tracks with stiffnesses in the tuned range were built at Harvard, Yale, Madison Square Garden, and the Meadowlands in New Jersey. The track at Harvard was designed with a stiffness of about 190 kN/m, enabling roughly 8 mm of track deflection for a 75 kg runner (assuming the runner exerts two times their body weight during midstance, per Figure 3.3). Not only did runners improve their times by about 2 or 3 percent (about 5 seconds per mile), but injury rates also decreased, perhaps because of the softer landing on the tuned track.

Today's tracks are built with attention to the amount of energy the track returns to the runner. No doubt their construction is more sophisticated than Harvard's track, which opened in 1977. Even if McMahon's work no longer represents the state of the art, his analysis of running and design of a tuned track are excellent

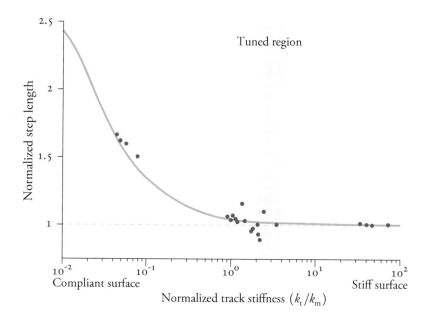

FIGURE 3.13
Step length (normalized by step length on a hard surface) vs. track stiffness (normalized by leg stiffness). Dots are experimental step lengths that are best fit by the curve. The tuned region (shaded) is the same as that shown in Figure 3.12. There is about a 1 percent increase in step length in this region. Adapted from McMahon (1984).

examples of just how much mileage you can get from simple models. Similar analyses can be used to design running shoes, as we will see next.

Elastic mechanisms to improve running shoes

Rodger Kram and his collaborators evaluated a prototype running shoe from Nike that included a carbon-fiber plate embedded in an elastic foam midsole, intended to store and release elastic energy. When a force is applied to a shoe, it deforms; analyzing the force–deformation relationship can tell us something about the storage and return of elastic energy. A comparison between the prototype shoe and a standard running shoe used by marathon runners revealed that the prototype shoe stored more elastic energy as it was deformed and returned a higher percentage of this energy when it recoiled (Figure 3.14). When 18 elite athletes ran in these shoes, their energetic cost of running decreased by an average of 4 percent compared with standard marathon racing shoes, similar to the advantage provided by the tuned track some four decades earlier.

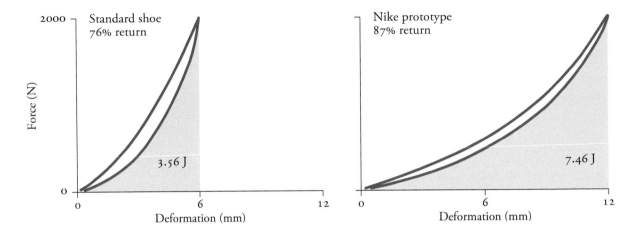

FIGURE 3.14
Force–deformation curves of two running shoes. As force is applied to each shoe, the shoe deforms (upper trace in each curve). Then, as the shoe is unloaded, the midsole recoils (lower trace in each curve). The area between loading and unloading curves indicates the mechanical energy lost as heat. The area below the lower traces (shaded) represents the amount of elastic energy that is returned when the shoe is unloaded. Adapted from Hoogkamer et al. (2018).

Leg stiffness changes with body mass

Analyses of mass–spring models have suggested that leg stiffness increases in proportion to body mass among a wide range of animals (Farley et al., 1993). For example, squirrels have a small body mass and move with flexed, compliant limbs whereas elephants have a large body mass and prefer to walk with straighter legs. Running while carrying load artificially increases body mass and is common in humans, making it an interesting scenario for exploring how mass influences leg stiffness during human running. When Amy Silder was working in my lab at Stanford, she collected data from recreational runners as each ran at his or her comfortable training pace (3.34 ± 0.22 m/s) on a treadmill that was instrumented with force plates. Subjects ran while carrying no load and while carrying loads of 10, 20, or 30 percent of their body weight in a vest. Dimensionless leg stiffness (k_{leg}) was estimated as the ratio of the peak vertical ground reaction force normalized by body weight (F_{peak}) to the change in leg length during that stance phase, normalized by leg length at foot contact (ℓ_0):

$$k_{leg} = \frac{F_{peak}}{(\ell_0 - \ell_{min})/\ell_0} \quad (3.4)$$

where ℓ_{min} is the minimum leg length during stance. Leg length was defined as shown in Figure 3.15.

These experiments revealed that dimensionless leg stiffness increased when running with load because the peak vertical ground reaction force increased and the change in stance-phase leg length decreased (Figure 3.16). These experiments agree with the trend noted by Farley et al. (1993) that leg stiffness increases with body mass.

Leg stiffness may also change with the type of running shoe worn. Running shoe cushioning is often used in an attempt to reduce impact loading, as we saw in the tuned track described above. However, Kulmala and colleagues (2018) reported that running in highly cushioned shoes increased leg stiffness and amplified impact loading compared to a conventional running shoe. Thus, one needs to consider the complex interaction of running shoe mechanics and limb stiffness when attempting to reduce musculoskeletal loading during running.

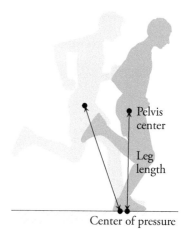

FIGURE 3.15
Leg length defined as the distance between the center of pressure and the geometric center of the pelvis.

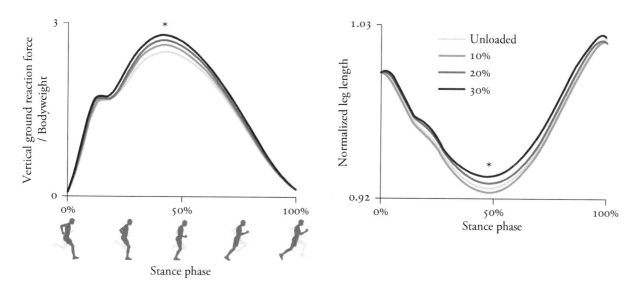

FIGURE 3.16
Vertical ground reaction force normalized by body weight (left) and leg length normalized by leg length at foot contact (right) over the stance phase. Subjects ran while carrying no load and while wearing vests loaded with 10, 20, or 30 percent of their body weight. As loads increased, the peak vertical ground force increased ($p < 0.001$) and change of leg length decreased ($p = 0.025$). Adapted from Silder et al. (2015).

Gait transitions

As walking speed increases, the stance phase shortens; ultimately, the double-support phase disappears entirely, producing a flight phase (Figure 3.17). It is interesting to examine what happens to the ground reaction forces during the transition from walking to

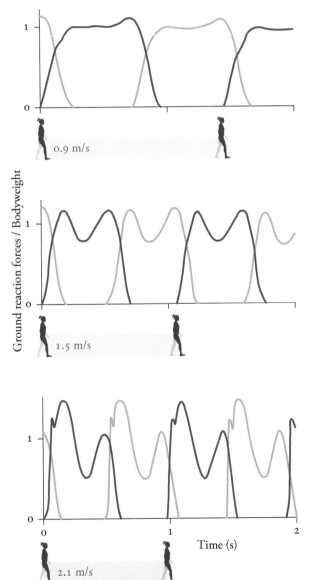

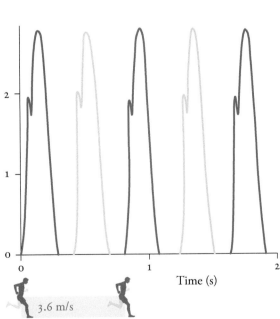

FIGURE 3.17
Vertical component of ground reaction forces as speed increases. Left, top to bottom: walking at slow, moderate, and fast speeds; right: running. Adapted from Alexander (1984).

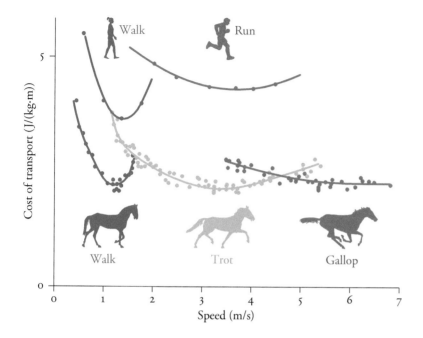

FIGURE 3.18
Cost of transport at various speeds. When moving at a particular speed, horses, humans, and other animals naturally select the gait with the lowest cost of transport, and prefer moving at speeds near the minima of these curves. Horse data from Hoyt and Taylor (1981); human data from Rathkey and Wall-Scheffler (2017).

running. At low walking speeds, the stance phase is long, as is the period of double support. As walking speed increases, the variation of the ground reaction force over the gait cycle increases and the period of double support drops. It's hard to walk at a speed of over 2 m/s; even though there is no flight phase, the peak ground reaction force increases to well above body weight and varies substantially over the gait cycle. At even higher speeds, it's impossible to walk and you break into a run. The peak ground reaction force increases dramatically, to over 2 bodyweights, because the foot is in contact with the ground for such a short duration. An important question to ask is: What drives this transition between gaits? Comparing the gaits of different animals provides important clues.

When moving at a particular speed, humans and other animals typically select the mode of locomotion with the lowest cost of transport. For example, horses transition from walking to trotting to galloping as they increase speed (Figure 3.18). Although they adapt their gait to the desired speed, the reverse is also true: once they have chosen a particular gait, they prefer to travel at the speed where the cost of transport is lowest for that gait. For example, horses typically walk at around 1.25 m/s, which is the lowest point of the blue curve in Figure 3.18. Humans similarly transition from walking to running around the speed at which running becomes more economical. Notice that, if you need to cover a particular distance in a fixed amount of time, it may be more energetically economical to walk comfortably part of the way and run the rest of the way rather than walking quickly or running slowly the whole way. As with walking, we naturally choose variables such as step length and step frequency to minimize energy consumption when running at a particular speed.

Bipedal mass–spring model

In Chapter 2 we modeled walking as the motion of an inverted pendulum with rigid legs, and in this chapter we have modeled running as a bouncing gait with compliant legs. Each model has strengths but also important limitations. In 2006, Hartmut Geyer, Andre Seyfarth, and Reinhard Blickhan proposed a unified model for walking and running that overcomes some of these limitations by including two compliant legs (Figure 3.19). The model reproduces three important features of running: the ground reaction force, elastic storage of energy, and a running gait cycle with alternating periods of stance and flight. Interestingly, the model also demonstrates the out-of-phase fluctuations in forward kinetic energy and gravitational potential energy that are characteristic of walking. Analysis of this model shows that walking efficiency depends on the center-of-mass trajectory following an inverted pendulum arc and on elastic energy storage in the leg springs.

FIGURE 3.19
An inverted pendulum model of walking on stiff legs (left) and a mass–spring model of running on compliant legs (center) can be combined into a bipedal mass–spring model that explains the basic dynamics of walking and running (right). The center-of-mass trajectory during walking is shown in the right panel. Adapted from Geyer et al. (2006).

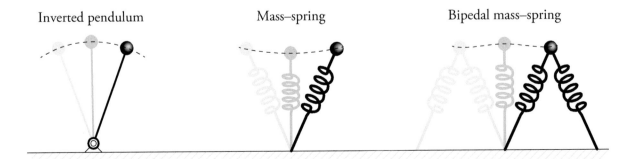

Energy that would otherwise be lost during double support is recovered from the energy stored in the leg springs. Having a single model that can produce both walking and running suggests that these gaits are more similar than was previously thought.

Kinematics of running

In addition to understanding the ground reaction forces and energy consumption involved in running, it is important to understand running kinematics. That is, we want to see how the joints pictured in Figure 2.17 behave: when they flex, when they extend, and by how much. Using motion capture methods, which we describe in Chapter 7, we can estimate the joint angles that occur during running and describe how joint motions vary with running speed. Typical results for subjects who run with a rearfoot-striking running style are shown in Figure 3.20. The ranges of joint motions are greater during running than during walking, and the ranges of motion increase with running speed.

The kinematics of the joints during the stance phase have several notable features. The hip is flexed at foot contact (0 percent of the gait cycle) and extends for the duration of the stance phase. In contrast, the knee begins in slight flexion at foot contact, flexes for the first part of stance, and then extends for the second part of stance. This knee flexion pattern contributes to the compression and extension of the leg during the stance phase, and thus to the spring-like behavior of the limb. For runners who land on their heels, the ankle is in a neutral position at foot contact; the ankle then dorsiflexes during the first part of stance and plantarflexes during the second part, working in concert with the knee to produce the spring-like behavior of the leg.

During the swing phase, the hip flexes to advance the limb. The knee flexes during the first half of swing so that the foot clears the ground, and extends during the second half of swing to prepare for landing. The ankle, which is highly plantarflexed at the end of stance, dorsiflexes back to a neutral position during the swing phase to prepare for landing.

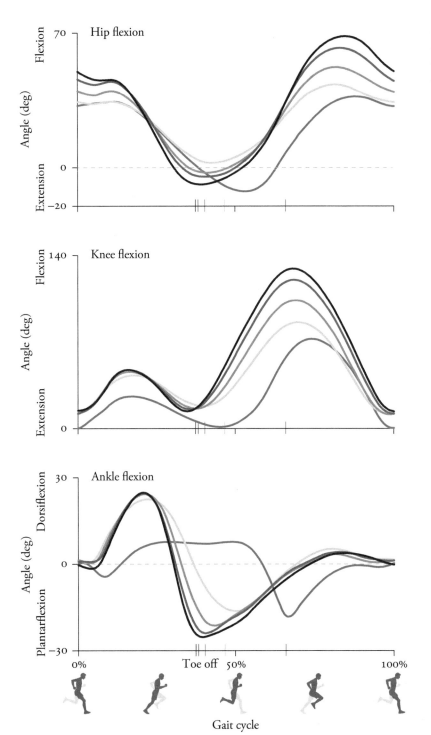

FIGURE 3.20
Representative joint motions over the gait cycle when running at several speeds. Walking at 1.5 m/s from Figure 2.19 is shown for reference. Averaged over 10 subjects. Vertical lines on the horizontal axis indicate toe off at each speed. Data from Hamner and Delp (2013).

Ground reaction forces and running speed

At low running speeds (e.g., 2 m/s), the vertical component of the ground reaction force peaks at about 2 bodyweights (Figure 3.21). As speed increases, so does the peak of the vertical ground reaction force, increasing to over 2.5 bodyweights at 5 m/s and up to 3 bodyweights at higher running speeds. Increasing speed also increases the horizontal force that is directed backward during the first half of stance, and the horizontal force that accelerates the center of mass forward during the second half of stance. The ground contact time shortens with increasing speed, which can be observed as a decrease in the fraction of the gait cycle spent in stance.

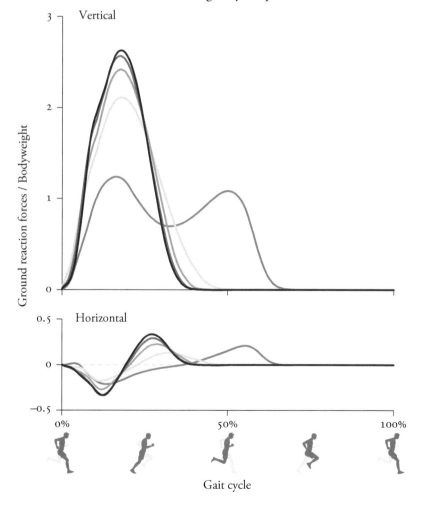

FIGURE 3.21
Representative ground reaction forces over the gait cycle when running at several speeds. Walking at 1.5 m/s from Figure 2.20 is shown for reference. Normalized by body weight and averaged over 10 subjects. Data from Hamner and Delp (2013).

So far, we have analyzed steady-state running at speeds up to 5 m/s, a fast marathon pace, but the fastest human sprinters can run more than twice this fast. Tim Dorn and colleagues studied high-speed sprinting and discovered that runners use two strategies to increase their speed (Dorn et al., 2012). Up to about 7 m/s, running speed is increased primarily by producing larger ground reaction forces to propel the body upward and forward, thereby increasing stride length. Above 7 m/s, the ground reaction force no longer increases, and the strategy to increase running speed shifts from increasing stride length to increasing stride frequency, achieved by moving the legs more rapidly during the flight phase. We will see how muscles are used to achieve high-speed running in Chapter 12.

In these two chapters on walking and running, we have used simple mechanical models to illustrate some of the fundamental features of gait, including how we move efficiently using pendular dynamics in walking and elastic energy in running. We have also examined ground reaction forces in detail because these recordings indicate how much force is applied to the foot and how the center of mass accelerates. We have gotten just about as far as possible without explicitly describing a crucially important ingredient in human locomotion—the muscles.

As we have hinted above, it is muscles and tendons that give rise to the spring-like properties of legs. Although we can measure ground reaction forces, we cannot determine which muscle forces produce those reaction forces without considering the dynamics of the musculoskeletal system. Once we calculate muscle forces, we will be able to estimate how large the loads are within the hip, knee, or ankle, and hence we can begin to understand joint injuries or design artificial joints. In the upcoming chapters, we will explore the structure, function, and computational modeling of muscle (Chapters 4–6), followed by the methods we use to estimate joint motions and muscle forces from experimental measurements of movement (Chapters 7–9). We will then be poised to explore how we coordinate our muscles during walking and running in Chapters 10–12.

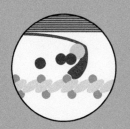 Part II
Production of Movement

4 Muscle Biology and Force

Alone we can do so little. Together we can do so much.
—Helen Keller

THE ROYAL SOCIETY OF LONDON is a 350-year-old institution with a nearly 200-year-old tradition of public lectures by scientists, often accompanied by simple experiments. In 1952, one of these experiments became the talk of the town.

Two stationary bicycles were oriented in opposite directions and linked together by a single chain, so that pedaling forward on one bike caused the pedals on the other bike to go backward (Figure 4.1).

FIGURE 4.1
The *push-me–pull-you* device described by Abbott et al. (1952).

On one bike sat a petite woman named Brenda Bigland, a leading expert on muscle fatigue. On the other sat a strapping young man named Murdoch Ritchie, who was married to his cycling "opponent." The lecturer was A. V. Hill, who had been awarded a Nobel Prize for his work on heat production in muscles.

On command, Ritchie pedaled forward as hard as he could, while Bigland resisted his pedaling as her pedals moved backward. Imagine the amazement of the audience when they saw the small woman easily preventing the large man from pedaling faster. He was soon sweating profusely and gasping for breath while she expended very little energy. Even the Lord Mayor of London came up afterward and wanted to give it a spin. "The equipment came to be named the *push-me–pull-you*, after Dr. Doolittle's two-headed animal that never knew which way it was going," wrote Brenda Bigland-Ritchie later.

Magic? Parlor trick? No, but it does show that there is nothing obvious about the way that muscles consume energy and generate forces. When muscles do positive work (such as when they act as the "motors" during forward pedaling) they consume more energy and generate more heat than when they do negative work (such as when they act as brakes to resist pedaling). In general, the forces produced and energy consumed by a muscle depend very much on whether it is shortening or lengthening. The lessons learned from Hill's experiment are still applied every day in rehabilitation and resistance training. As we will see, the magic happens at the molecular level.

This chapter and the next take a deep look inside muscles, exploring the relationship between their structure and function. Muscles are amazing biological motors, capable of silently generating thousands of newtons of force in a fraction of a second. The forces are so large that the muscles of your calf could pick up the back end of a small car. These large forces are generated through the combined actions of trillions of nanoscale molecular motors that convert the chemical energy derived from the food we eat into mechanical work that enables us to move. This remarkable function is a result of the specialized cellular machinery and the highly organized hierarchical structure of skeletal muscle (Figure 4.2).

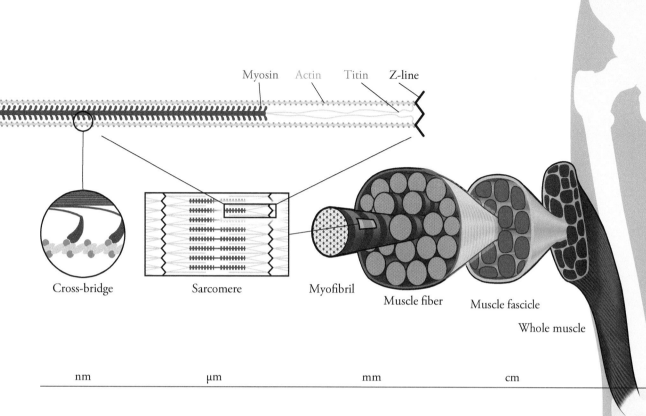

FIGURE 4.2
Multiscale structure of muscle. Skeletal muscle is structured hierarchically, with nanoscale molecular motors called myosin, each of which generates only a few piconewtons of force, arranged into sarcomeres, myofibrils, fibers, fascicles, and whole muscles that can generate thousands of newtons of force during powerful muscle contractions.

The organization of the next two chapters will roughly follow the hierarchical structure of muscle. We will begin by studying the force-generating process at the molecular level. Next, we will see how molecular motors are packed into subcellular structures called sarcomeres. (The first image of sarcomeres in a live human was captured in my arm and is shown in this chapter's opening figure.) Moving up in scale, we will see how individual muscle cells are activated by the nervous system and learn the difference between "fast-twitch" and "slow-twitch" muscle fibers. In Chapter 5, we will look at the macroscopic scale and see how muscles are arranged in the body and how they interact with tendons, the structures that convey muscle forces to the bones.

Muscle structure

At the most basic level, muscles generate force via the interaction of two long, narrow proteins called actin and myosin. In the early 1950s, Hugh Huxley discovered through X-ray microscopy that these proteins line up in parallel, interlacing fibers, with connections between them that he called *cross-bridges*. Andrew Huxley (no relation to Hugh) discovered cross-bridges at the same time using different methods. Both Huxleys suspected that the cross-bridges were the mechanism that generated force, and in 1954 they proposed a model for how these molecular motors work, which we explain below. Their model has remained the basic paradigm for explaining force generation ever since, although it has become more detailed as more experimental data have been collected.

Any single molecular motor generates only a few piconewtons of force when forming a cross-bridge. Massive teams of these motors are organized into subcellular structures called *sarcomeres*, which are about 3 μm long. These in turn are connected end-to-end to form long, thin strands called *myofibrils*. In long muscles, a myofibril may contain more than 50,000 sarcomeres in series. Myofibrils are arranged in parallel to form a *muscle fiber* or myocyte ("muscle cell"). Muscle fibers are roughly 50 μm across and can be tens of centimeters long. Because muscle fibers are cells, they contain the typical organelles required for essential cellular functions, such as nuclei and mitochondria. They also contain a special-purpose structure called the sarcoplasmic reticulum, which stores the calcium required to activate muscle. We will explore the role of calcium in more detail later in this chapter.

Moving up the size scale, muscle fibers are arranged in bundles called *muscle fascicles* which, like muscle fibers, can be tens of centimeters long. The cross-section of a muscle fascicle measures about 1 mm across. In addition to muscle fibers, the fascicle also contains connective tissue called the extracellular matrix, which includes collagen, nerve fibers, and blood vessels. Muscle fibers are packed tightly in healthy muscle; in diseased muscle, however, muscle fiber cross-sections can be smaller and separated by more extracellular matrix and fat.

The fascicles are surrounded by more connective tissue, and pack together to form a muscle. An additional sheath of connective tissue called fascia encapsulates the muscle and separates it from other muscles. The muscle fibers that terminate at the end of each fascicle may attach directly to bones, but typically they connect to tendons which then attach to bones. The region of the tendon that inserts into the muscle is known as the *aponeurosis*; the region of tendon that is external to the muscle is often called the *free tendon*. As we will see in Chapter 5, tendons play important roles by not only transferring muscle forces to the skeleton, but also storing and releasing energy as they stretch and recoil.

The cross-bridge cycle

When activated by the nervous system, muscles generate force through the coordinated action of trillions of actin and myosin proteins in a process called the *cross-bridge cycle* (Figure 4.3). This mechanism can be roughly described as a molecular-sized ratchet.

The myosin molecule has three regions, known as the head, neck, and tail. The unique structure of the myosin head allows it to bind firmly to certain locations on the actin filament (first frame of Figure 4.3). When the muscle is called on to produce force, the myosin head receives exactly one molecule of a fuel called adenosine triphosphate (ATP). This stimulates the myosin head to detach from the actin, rotate forward, and attach to the next binding site on the actin (frames 2 and 3 in Figure 4.3). Next, the myosin head pivots about the neck region in a motion referred to as the *power stroke*, which generates a few piconewtons of force and slides the filaments about 10 nm past each other.

FIGURE 4.3
The cross-bridge cycle describes the process by which actin and myosin interact to produce force and motion. A is adenosine; P is phosphate.

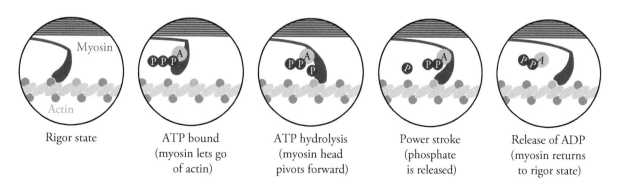

Of course, it is impossible to perform mechanical work without consuming energy. The energy comes from a chemical reaction that detaches one phosphate ion from the ATP (frame 4 in Figure 4.3), leaving adenosine diphosphate (ADP). Finally, the ADP is released and the myosin returns to its original state, but now displaced from its original position. As myosin proteins cycle in this way, the thin filaments are pulled toward the middle of the sarcomere, thereby creating tension as chemical energy is converted into mechanical energy.

While this mechanism might sound complex, it is an elegant and economical solution to the problem of how to generate force in a predictable direction at the molecular level. Like automobile engines, our muscles convert chemical energy into mechanical work, but with less noise and less toxic exhaust.

Sarcomere structure

Moving one step up in scale, we come to sarcomeres, which are roughly cylindrical and vary in length between about 1 and 5 μm depending on the length of the muscle. Sarcomeres are comprised of many interlacing "thick" and "thin" filaments, which slide past each other as the sarcomere changes length.

Hundreds of myosin tails are bundled together to form each thick filament, giving rise to rodlike structures with myosin heads emanating radially outward at regular intervals (Figure 4.4). Running parallel to the thick filaments are the thin filaments, which are composed of three proteins: actin, tropomyosin, and troponin. We have already described actin, which provides binding sites for the myosin heads. The binding sites are spaced at regular intervals along the thin filament. Tropomyosin and troponin help to regulate force production by exposing these binding sites only when the muscle is activated by the nervous system and calcium is present. One additional protein of interest, called titin, attaches each thick filament to the ends of the sarcomere (which we call the Z-lines or Z-discs). As we will see later, titin plays an important role in passive force generation, a separate process from the

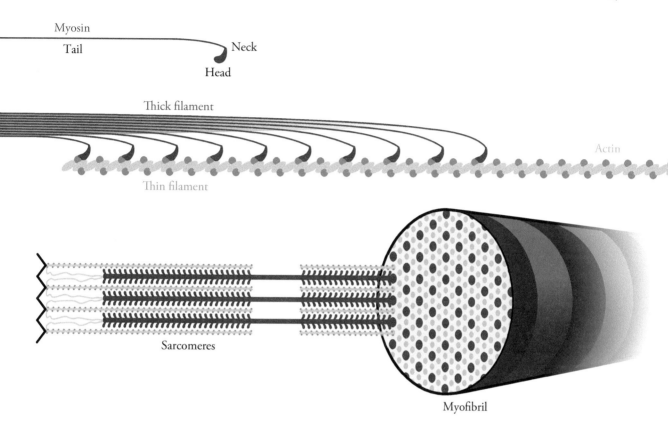

FIGURE 4.4
Schematics of myosin (top), interaction of thick and thin filaments (center), and cross-section of myofibril showing thick and thin filaments arranged in a highly organized three-dimensional pattern (bottom).

cross-bridge cycle and a phenomenon that the Huxleys' original model missed.

Sarcomeres are arranged in series and parallel within skeletal muscle in a regular pattern, resulting in striations, or alternating light and dark bands, when skeletal muscle tissue is viewed under a microscope. You can see these bands in Figure 4.5, which go by the names of I-bands, A-bands, Z-discs, and M-discs. Properly speaking, the Z-discs and M-discs are indeed discs, but they are often called Z-lines and M-lines because that is what they look like when viewed in two dimensions. The A-bands appear only in regions containing myosin, and the I-bands appear elsewhere. The thin actin filaments are anchored at one end to the Z-discs. The thick myosin filaments attach at one end to structures in the M-disc and at the other end are tethered to the Z-discs by titin molecules. Skeletal muscles are sometimes referred to as "striated" muscles—in contrast to "smooth" muscles, such as those that control the caliber of blood vessels, which are not organized into sarcomeres and do not exhibit striated patterns.

Force–length relationship

The maximum force that can be generated by a sarcomere changes as the sarcomere changes length. The relationship between length and tension can be explained by the *sliding filament theory*, which was independently proposed by two research teams in 1954. Andrew Huxley and Rolf Niedergerke (in single muscle fibers) and Hugh Huxley and Jean Hanson (in isolated myofibrils) showed that the A-band does not narrow during active muscle contraction, which disproved the prevailing theory that the thick filaments shorten. The explanation they offered was that the thick and thin filaments slide past each other as the sarcomere changes length. As a consequence, the length of the sarcomere affects the amount of "overlap" between the actin and myosin, or the number of myosin heads that are in proximity to binding sites on the thin filament. As the number of cross-bridges between myosin heads and actin binding sites increases, so does the tension within an activated sarcomere.

The force that can be generated by cross-bridge cycling in activated muscle varies with sarcomere length as shown in Figure 4.6. This *active force–length curve* is often described as having three regions: an ascending region where force increases with increasing sarcomere length, a plateau region where force remains at its maximum, and a descending region where force decreases with increasing sarcomere length. The plateau region spans a range of sarcomere lengths known as the "optimal" range, where the number of interactions between myosin heads and actin binding sites is at a maximum. The optimal sarcomere length varies across vertebrates, but generally lies between 2.2 and 2.7 μm, and is around 2.7 μm in human skeletal muscle. The amount of actin–myosin overlap decreases when sarcomeres become longer than the optimal range, as does the force that can be generated by cross-bridge cycling. At

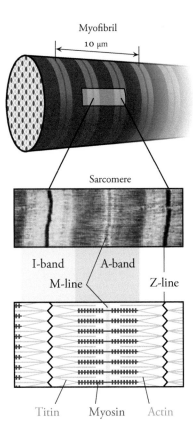

FIGURE 4.5
A schematic of a myofibril (top) shows its highly organized microscopic structure. The I-band, A-band, M-line, and Z-line are labeled in a microscopic image of muscle (middle) and a schematic of a sarcomere (bottom). The ends of the sarcomere are defined by the Z-lines.

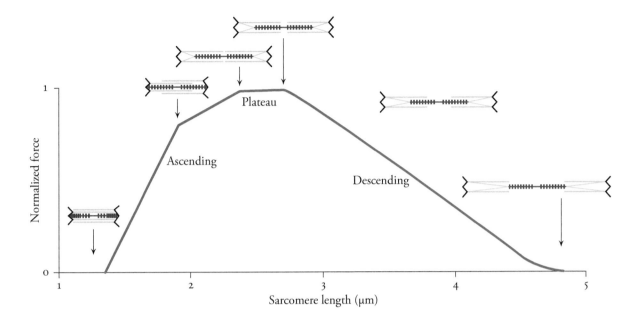

FIGURE 4.6
The active force generated by a sarcomere is a function of its length. Force varies as the thick and thin filaments slide past each other. Force increases with length in the ascending limb of the curve, plateaus, and decreases with length in the descending limb. Force generation peaks when the maximum number of cross-bridges are formed. Adapted from Gordon et al. (1966).

lengths slightly shorter than optimum, the thin filaments originating at opposite ends of the sarcomere begin to overlap and interfere with each other in the middle of the sarcomere, hence the decrease in force in this so-called "shallow ascending region." At even shorter lengths (in the "steep ascending region"), the thick filaments collide with the Z-discs and deform, generating a force that opposes the action of the cross-bridge cycling. Because muscles are composed of sarcomeres arranged in series and parallel, we can observe a similar relationship between force and length during active contraction in whole muscle.

Muscles generate force when stretched beyond their resting length even when they are inactive, like a nonlinear spring. This *passive force–length relationship* arises primarily from two structures in the muscle: titin and the connective tissue surrounding fibers, fascicles, and muscle. Titin is a large, flexible protein that tethers the thick filament to the Z-discs at either end of the sarcomere. Titin is coiled and has low stiffness when the sarcomere is short; as the sarcomere lengthens, the titin straightens and its stiffness increases (Figure 4.7). The connective tissue that surrounds muscle fibers exhibits a similar force–length behavior. The total force generated by the muscle when it is activated is the sum of its passive force and the active force generated by the cross-bridges (Figure 4.8).

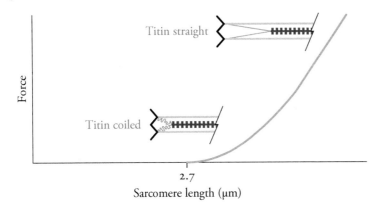

FIGURE 4.7
Titin attaches the thick filaments to the Z-discs at either end of the sarcomere and develops force when stretched.

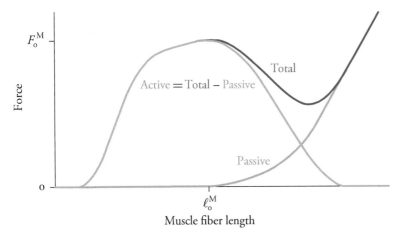

FIGURE 4.8
The active force–length curve can be determined by subtracting measurements of the passive force from the total force over a range of fiber lengths. The peak active force, F_o^M, occurs at the optimal muscle fiber length, ℓ_o^M.

Force–velocity relationship

It may come as some surprise that the force generated by muscle depends not only on its length but also on the rate at which the length is changing. We characterize this rate dependence with the *force–velocity curve* (Figure 4.9). This relationship was first observed by A. V. Hill in 1938, and it eventually led him and his colleagues to develop the "push-me–pull-you" experiment described at the beginning of this chapter.

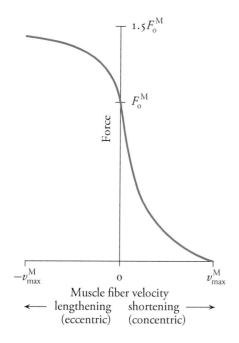
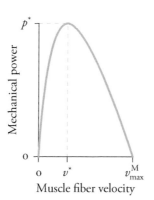

FIGURE 4.9
Muscle fiber force and power as functions of the fiber's velocity. Force-generating capacity increases when lengthening and decreases when shortening (left). Mechanical power, the product of force and velocity, reaches a maximum (p^*) at about one-third of the maximum shortening velocity (v^*; right).

Here is why force depends on velocity. Sarcomeres generate force by way of the cross-bridge cycle, in which myosin heads bind with and pull on the actin filament. Because this process takes time, the active force that can be generated through cross-bridge cycling decreases when the sarcomere is shortening. At very high shortening velocities, no active force can be generated because the thick and thin filaments are sliding past each other faster than the myosin heads can pivot during the power stroke. Conversely, the active force increases when the muscle is lengthening because the neck regions of the myosin are strained before the heads detach from the thin filaments. Very high lengthening velocities can damage these and other structures, resulting in dramatic decreases in active muscle force.

A muscle's velocity is of great significance when exercising. In an *isometric* exercise, a muscle is activated while pulling on an equal resistive force. In this case, the muscle will not change length (i.e., it will have zero velocity). If a muscle is opposed by a weaker resistive force, the muscle will shorten and its contraction is said to be *concentric*. Conversely, a muscle will elongate if it is opposed by a greater force, in which case its contraction is said to be *eccentric*.

Hill observed that the active force decreases hyperbolically as shortening velocity increases during concentric contractions, reaching zero at the *maximum contraction velocity* (v_{max}^M). During eccentric contractions, the muscle force increases beyond its force during an isometric contraction, reaching up to around 1.5 times its maximum isometric force.

Hill's observations explain why Murdoch Ritchie had to work so hard during the "push-me–pull-you" experiment, while Brenda Bigland scarcely broke a sweat. The apparatus forced Ritchie's muscles to operate on the "concentric" side of Figure 4.9, where they exerted far less than their maximum force. By contrast, Bigland's muscles were operating on the "eccentric" side of the force–velocity curve, where they could exert nearly their maximum force. Ritchie, pedaling forward, was acting as the motor, while Bigland, pedaling backward, was acting as a brake.

Another distinction between concentric and eccentric contraction is the nature of the work done by the muscles. Muscles do positive work during concentric contraction because force and displacement are in the same direction. During eccentric contraction, however, muscles have work done on them (i.e., muscle work is negative) because the force and displacement are in opposite directions.

You may wonder whether muscles perform both positive and negative work in daily life. They do! When cycling or walking uphill, your muscles are performing primarily positive work. When walking down a steep slope, however, many of your muscles are doing negative work to control the speed of your descent. Interestingly, doing negative work makes muscles sore and promotes muscle regeneration.

Negative work suggests that, in principle, muscles can gain energy or "recharge" themselves from an external force. Alas, this does not happen. Our bodies are not equipped to reverse the chemical reaction that powers the cross-bridge mechanism and convert ADP (adenosine diphosphate) back into ATP (adenosine triphosphate). However, recharging is possible for machines powered by rechargeable batteries. For example, whereas the friction brakes in conventional vehicles convert kinetic energy into heat, electric automobiles use regenerative braking to capture and store this energy.

The force–velocity relationship has an important consequence for the mechanical power generated by contracting muscle (Figure 4.9, right panel). When the muscle is isometric, large forces can be generated but the velocity is zero; thus, the mechanical power, which is the product of force and velocity, is zero. The mechanical power is also zero at and above shortening velocities of v_{max}^M, where the velocity is large but no force can be generated. The peak positive mechanical power occurs at approximately $\frac{1}{3} v_{max}^M$, when the product of force and velocity is at a maximum. The power–velocity relationship does well at predicting some features of everyday movement. For example, cyclists shift gears when they sense their muscles moving too fast or too slow, venturing far from the range of maximum power.

Muscle activation

We have assumed that actin and myosin can engage in cross-bridge cycling as long as the myosin heads are in proximity to binding sites on the thin filament and ATP is present. However, when a muscle is relaxed, the myosin-binding sites on actin are blocked by the protein tropomyosin. When a muscle is excited, calcium ions are released into the intracellular space and bind to the protein complex troponin, which causes the tropomyosin to change shape and expose the binding sites on the thin filament (Figure 4.10).

It's natural to wonder: Where do the calcium ions come from? As mentioned earlier, they are stored in a sort of holding tank in

FIGURE 4.10
Molecular changes during muscle activation. When a muscle is relaxed (left), the myosin-binding sites on the actin filament are blocked by the tropomyosin protein. Activating a muscle (right) causes a release of calcium ions from the sarcoplasmic reticulum. The calcium ions bind to the troponin protein complex, causing the tropomyosin to change shape and reveal the myosin-binding sites on the actin, thereby allowing cross-bridge cycling to begin and force to be generated.

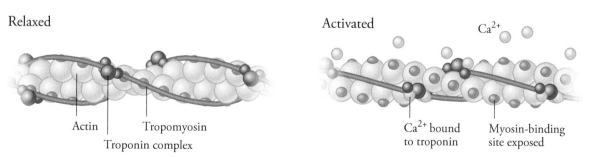

the cell. Their release is orchestrated by the central nervous system (CNS), an extremely important player in the muscular story that we have not mentioned yet. Muscles do not activate themselves; under ordinary circumstances, they become active only when instructed to do so by the nervous system.

The muscle fiber has several structural features that allow it to receive signals from the CNS and quickly propagate those signals to release calcium ions along the entire length of the cell. A motor neuron joins with each fiber at a specialized synapse called a *neuromuscular junction* (Figure 4.11). An excitatory signal from the CNS results in the transfer of a neurotransmitter called acetylcholine across the neuromuscular junction synapse. This transfer triggers depolarization of the cell membrane and causes an outflow of positive calcium ions from the sarcoplasmic reticulum. The transverse-axial tubular system (often referred to simply as the T-tubules) ensures that the depolarization propagates quickly along the entire length of the muscle fiber. The T-tubules are formed by the cell membrane (called the sarcolemma) as it dips into the cell, extending membranous fingers into the cytoplasm. Located adjacent to the T-tubules are the terminal cisternae, extensions

FIGURE 4.11
The structure of a muscle fiber enables rapid propagation of action potentials and coordination of cross-bridge cycling along its entire length.

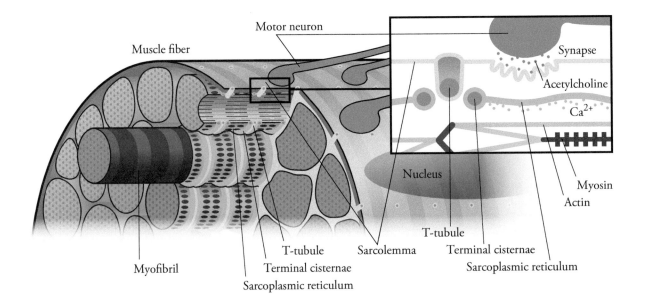

of the sarcoplasmic reticulum that store calcium. When the cell depolarizes, this calcium is released into the intracellular space and initiates cross-bridge cycling.

When the excitatory signal from the CNS ends, the cell begins pumping calcium ions back into the sarcoplasmic reticulum. This calcium reuptake process is slower than the release process because calcium ions are being transported against a concentration gradient.

So far, we have examined only how the central nervous system activates a muscle. But its job is not done! It also needs to tell muscles whether to pull hard or gently. The CNS uses the above machinery to modulate the amount of force generated by a muscle in two ways: rate encoding and motor unit recruitment. We describe these mechanisms in the next two sections.

Rate encoding

Information is coded and transmitted along a motor neuron in the form of electrical impulses called *action potentials*. Ordinarily, the signal to activate comes from the brain, but a muscle can also be stimulated artificially by passing electric current through its innervating motor neuron. Functional electrical stimulation therapy exploits this property by electrically stimulating peripheral motor neurons in individuals with paralysis due to spinal cord injury. Functional electrical stimulation systems can be used to strengthen muscles and produce coordinated movements such as rowing, grasping, and cycling, as we saw in Chapter 1.

A short electrical stimulus applied to a motor neuron causes a "twitch" in the muscle fibers it innervates (Figure 4.12), provided the stimulus exceeds a threshold voltage. There is a small delay between application of an electrical impulse and force production as the depolarization propagates and calcium ions are released. The fiber force then rises rapidly, peaks, and falls back to zero relatively slowly as the reuptake process pumps calcium ions out of the intracellular space.

Muscle fibers are generally not stimulated to produce a single twitch; rather, a train or sequence of impulses is applied at a

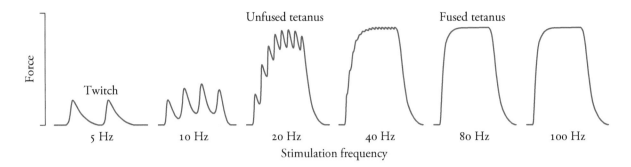

FIGURE 4.12
Muscle force resulting from different stimulation frequencies. Fiber force is modulated by the frequency of stimulation (rate encoding), producing force responses ranging from a twitch (left) to fused tetanus (right).

particular rate. If two impulses are applied in sufficiently close succession, the second twitch will begin before the force from the first has fully decayed to zero, resulting in a second force peak that is higher than the first. As the frequency of an impulse train (or the "firing rate") increases, so too does the peak force—until a threshold stimulation rate has been reached where the fiber force plateaus and the peak force does not increase at higher frequencies. This threshold stimulation rate is called the *tetanic frequency*. A muscle operating at or above the tetanic frequency is said to be in *fused tetanus* (the discrete twitches have fused together to form a smooth force profile); at lower frequencies, the muscle is said to be in *unfused tetanus*. Muscle force can be modulated approximately fourfold from twitches to tetanus by varying the stimulation frequency. In other words, the modulation of force can be expressed in the rate of stimulation, hence the term *rate encoding*.

Of course, many movements would be impossible if rate encoding were the only mechanism for modulating force production. Many muscles require greater than a fourfold variation in force to perform the range of functions for which they are used, and our movements would be shaky if submaximal muscle forces fluctuated like the ones shown in Figure 4.12.

Motor unit recruitment

As might be expected, the second mechanism for controlling muscle force production involves modulating the amplitude of the control signal rather than its frequency. The nervous system accomplishes this by recruiting only some of the muscle fibers available for any given task: fewer to generate small amounts of force, more to generate larger forces (Figure 4.13). During submaximal efforts, some muscle fibers are left inactive.

In more detail, this is how the process works. Muscles are functionally separated into distinct *motor units* that consist of one motor neuron and the muscle fibers it innervates (Figure 4.13, left panel). Motor units vary in size from tens of muscle fibers to thousands. Small muscles that require fine control, such as muscles of the thumb, have only a few muscle fibers per motor neuron; larger muscles that can be controlled more coarsely, such as muscles of the back, have many muscle fibers per motor neuron. Typically, the fibers belonging to a motor unit are not clustered together, but rather are distributed throughout the muscle. Thus, adjacent fibers generally belong to different motor units, and the fibers innervated by a particular motor neuron may be located in several fascicles.

FIGURE 4.13
Motor unit recruitment. A motor unit comprises a motor neuron and the muscle fibers that it innervates. Three motor units (blue, orange, and red) are shown (left); the central nervous system modulates muscle force by adjusting the number of motor units being recruited (right).

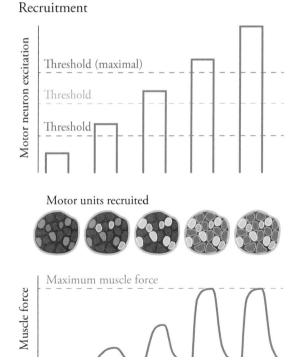

The CNS adjusts the number of motor units that are recruited based on the demands of the task (Figure 4.13, right panel).

Motor units are recruited in a particular order, referred to as *orderly recruitment*, as described by *Henneman's size principle* (Henneman et al., 1965). To generate small amounts of force, the CNS recruits only small motor units. As the required muscle force increases, the CNS recruits progressively larger motor units as well. Small and large motor units differ in more than just size: the small motor units that are recruited preferentially are also more resistant to fatigue. Thus, when performing low-intensity movements, the CNS selectively recruits small fibers so as to avoid muscle fatigue. The significance of orderly recruitment will be discussed in more detail in Chapter 5.

In contrast to orderly motor unit recruitment, externally applied electrical stimulation does not recruit motor units according to Henneman's size principle. When muscles are stimulated externally, large motor units are recruited at the same time as small ones (Gregory and Bickel, 2005). This has implications for functional electrical stimulation. In particular, recruiting large, fatigable fibers at low levels of stimulation makes it difficult to modulate force precisely and results in rapid fatigue.

Clearly, the quality of therapy could be improved if we could induce orderly recruitment in individuals with paralysis. Recent experiments in my laboratory have demonstrated that orderly recruitment can be achieved artificially using optical control to activate motor neurons that have been altered to respond to light (Llewellyn et al., 2010). This technology may eventually lead to more effective therapies and treatments for individuals with spinal cord injury and other motor impairments.

Electromyography

When muscle is excited by the nervous system, it generates a small electrical signal. We just saw that the nervous system modulates the amount of force generated by muscle via rate encoding and motor unit recruitment. Roughly speaking, the greater the stimulation arriving to a muscle via the motor neurons (either from a higher frequency or greater recruitment), the larger the electrical signal will be. Thus we can measure the level of muscle activity using *electromyography* (EMG), which, as its name suggests, records electrical activity in the muscle. If we connect EMG electrodes to an amplifier and speaker, we can literally hear the muscles working. Since we will frequently discuss research that uses EMG signals in this book, it is important to understand the basics of how we make these measurements.

To record muscle activity, we either mount electrodes to the skin for "surface" EMG or insert wires into muscle for intramuscular EMG. Surface electrodes are used to record the activity of large superficial muscles, whereas intramuscular electrodes are best suited to smaller or deeper muscles. In most setups, a single pair of electrodes (reference and signal) is used for each muscle; thus, the measured time series represents a combination of the firing of all recruited motor units within the muscle—and, potentially, surrounding muscles. To complicate matters further, the magnitude of the measurement is sensitive to electrode placement relative to the nerve and muscle fibers (which may be difficult to control precisely, particularly during movement). Consequently, the raw EMG signal is noisy and can be challenging to interpret.

A measured EMG signal is typically filtered and otherwise processed to increase its interpretability (Figure 4.14). We typically high-pass-filter the EMG signal to remove drift over time, then rectify and low-pass-filter it to obtain an envelope that indicates how the magnitude of the muscle's activity varies over time. Finally, we normalize the EMG signal relative to a calibration measurement collected during a maximum voluntary

FIGURE 4.14
Processing of electromyographic (EMG) signals. Raw EMG signals increase in amplitude as motor unit firing rate increases and as more motor units are recruited. To facilitate interpretation, a raw EMG signal is typically processed by high-pass filtering, rectifying, low-pass filtering, and then normalizing to a maximal signal.

contraction or a high-intensity dynamic movement. The result is a representation of the muscle's activity as a percentage of its maximum.

Two delays are often introduced in the processing of an EMG signal to relate the signal to a muscle's force production. First, there is a delay between the appearance of a nonzero EMG signal and a nonzero muscle force; the duration of the delay may vary with electrode placement and between muscles. A second delay is due to activation dynamics, as discussed below. EMG provides a gross measurement of muscle activation, including information from both rate encoding and recruitment. When properly calibrated, EMG can be a useful tool for studying muscle coordination and for testing the accuracy of simulations involving muscle activations.

Modeling muscle activation dynamics

Now that we have covered some fundamentals of muscle biology, we can form engineering models to describe muscle mechanics. To do this, we need to characterize the transformation from muscle excitation to muscle activation, which we call activation dynamics, and also develop equations that represent the contraction dynamics of muscle (Figure 4.15). We consider activation dynamics below; contraction dynamics will be covered in the next chapter.

The effects of rate encoding and motor unit recruitment can be incorporated into a mathematical model of activation dynamics. There are many strategies for modeling muscle activation; my

FIGURE 4.15
Inputs and outputs of a muscle–tendon model. In computational models of muscle–tendon actuators, we assume that the process of activation dynamics, which converts excitation ($u(t)$) to activation ($a(t)$), is distinct from contraction dynamics, which relates muscle–tendon length ($\ell^{MT}(t)$) and velocity ($v^{MT}(t)$) to muscle force ($F^M(t)$).

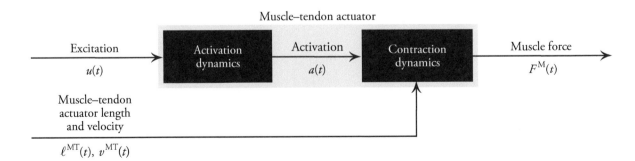

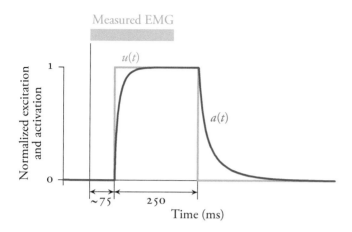

FIGURE 4.16
A computational model of activation dynamics relates excitation ($u(t)$) to activation ($a(t)$) using a first-order ordinary differential equation. Excitation often lags behind a measured EMG signal, and force production lags further still.

laboratory uses the following approach. The model accepts as input a time-varying function $u(t)$ that describes the strength of the *excitation* signal from the nerve to the muscle. Its output, $a(t)$, is the *activation*, which represents the availability of calcium ions within the intracellular space and, thus, the extent to which cross-bridge cycling can occur. Both $u(t)$ and $a(t)$ vary between 0 (no excitation; no cross-bridge cycling) and 1 (maximum excitation; fused tetanus and all motor units recruited).

We model the relationship between excitation and activation with an empirically determined first-order ordinary differential equation:

$$\dot{a}(t) = \frac{u(t) - a(t)}{\tau}$$

$$\text{where } \tau = \begin{cases} \tau_A(0.5 + 1.5a(t)), & \text{if } u(t) > a(t) \\ \dfrac{\tau_D}{0.5 + 1.5a(t)}, & \text{otherwise} \end{cases} \quad (4.1)$$

Parameters τ_A and τ_D represent the activation and deactivation time constants. As mentioned above, τ_A is smaller than τ_D; typical values are 10 and 40 ms, respectively, but these values may vary with age, muscle composition, and other factors. Figure 4.16 shows a typical solution to this differential equation, when the input $u(t)$ is a square wave.

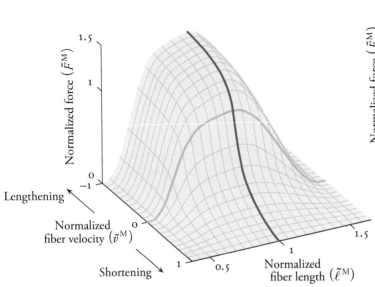

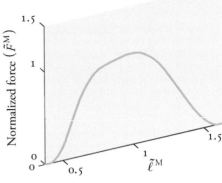
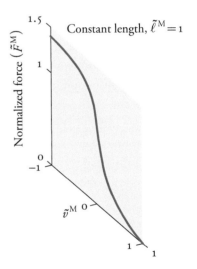

FIGURE 4.17
Muscle force-generating capacity varies with fiber length and velocity. Adapted from Lieber (2010).

Modeling the force–length–velocity–activation relationship

As we have seen, the force generated by a muscle depends on the length and velocity of its fibers. Muscle force is also modulated by motor unit firing rate and recruitment. The functional relationship between force, length, and velocity can be illustrated at maximal activation ($a(t)=1$) as shown in Figure 4.17, where the force-generating capacity at a particular length and velocity is simply the product of the corresponding values on the force–length and force–velocity curves. Notice that the cross-sections of this graph created by holding one variable constant (either length or velocity), shown in blue and green, are exactly the same as the active force–length and force–velocity curves we have already seen. In Figure 4.18, we see the effect of submaximal activations ($a(t)<1$), which scale the surface plot downward. The passive component of the force–length relationship is not illustrated in Figures 4.17 or 4.18, but it is always contributing force in healthy muscle and does not depend on activation.

The model shown in Figure 4.18 is, of course, a simplification of how real muscle behaves. For example, we have assumed that fiber length, fiber velocity, and activation affect muscle force generation independently. In real muscle, these relationships are more complex. Also, force generation depends on past states as well as temperature and fatigue. Nevertheless, as we will see, simplified models like these are powerful tools because they are easy to understand but sufficiently detailed to provide insight into muscle-driven movement.

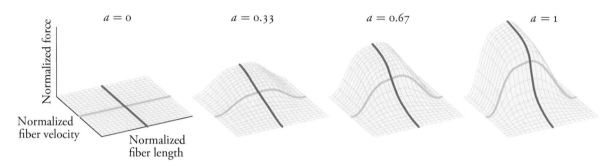

We all know that muscles come in a wide variety of shapes and sizes. In the next chapter, we will see how to use the concepts we have covered here to formulate a computational model of muscle—that is, a model that can be used to perform numerical analyses and generate simulations of movement. The model will enable us to represent the force-generating properties of many different muscles in a wide variety of conditions.

FIGURE 4.18
The nervous system modulates muscle force through rate encoding and motor unit recruitment, collectively modeled by muscle activation (a).

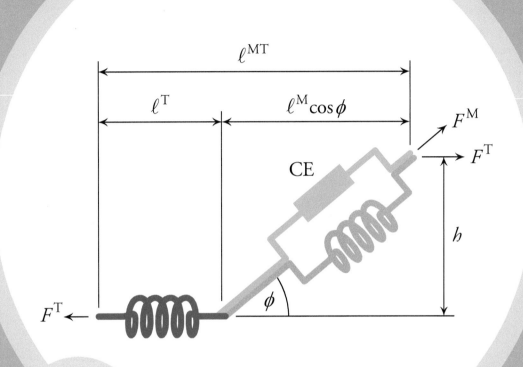

5 Muscle Architecture and Dynamics

If you want to understand function, study structure.
—Francis Crick

RESEARCHERS SEEKING TO UNDERSTAND human and animal movement have performed a wide variety of experiments. Biomechanists have studied whole-body movement by measuring joint motions, ground reaction forces, and electromyographic signals from thousands of people. Physiologists have studied isolated muscle to characterize the dynamics of muscle activation and force production. Muscle-driven simulations allow us to connect these two worlds, synthesizing biomechanical measurements of whole-body movement with experiments performed on individual muscles. We will see in Chapters 10–12 that muscle-driven simulations provide insight into the role of muscles in producing movement and provide estimates of important quantities that are nearly impossible to measure in people as they move, such as the force generated by a muscle and the amount of energy it consumes.

Modeling the dynamics of muscles is necessary for creating muscle-driven simulations of movement. However, a one-size-fits-all model is not adequate because each muscle has a distinct structure to suit its unique functions. For example, some muscles provide fine motor control of the fingers while others support body weight during locomotion (Figure 5.1). All skeletal muscles have a hierarchical arrangement of sarcomeres, but muscles differ in several important respects. Among these differences are their size and *architecture*, the geometric arrangement of their fibers. A computational model

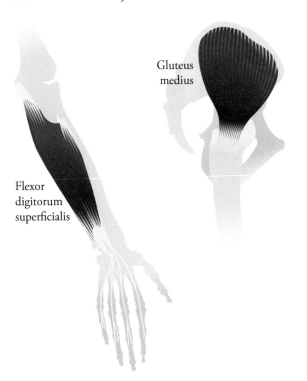

of muscle must, therefore, capture the features of muscle force generation that are common to all muscles while also being able to represent the unique features of each muscle.

In this chapter, we will see how to create a generic model of muscle force production and how to customize it to represent nearly any muscle in the body. The muscle model we will describe belongs to a class of models named after A. V. Hill, who performed many fundamental studies of muscle in addition to the "push-me–pull-you" experiment we saw in Chapter 4. Felix Zajac, my doctoral advisor, enhanced the Hill-type model and brought it into the modern era of computer simulation (Zajac, 1989). In particular, Zajac developed a model that includes only four generic curves and five muscle-specific parameters, all of which can be derived from experimental data and used to tune the model. The simplicity of Zajac's model is crucial for dynamic simulations that involve dozens of muscles, yet it is sufficiently detailed to represent the dynamics of muscles of different size, strength, and structure.

FIGURE 5.1
Muscle architecture and function vary throughout the body. The flexor digitorum superficialis (far left) flexes the fingers via four tendons; the broad gluteus medius and thin gracilis generate hip abduction and adduction moments; the gastrocnemius (far right) inserts into the calcaneus via the long Achilles tendon.

The features of muscle force generation common to all muscles are summarized in Figure 4.18. These features include three curves that describe the nonlinear relationships between a muscle's length and the force it generates: the active force–length curve, the passive force–length curve, and the force–velocity curve. Because muscle attaches to bone via tendon, we must also consider the properties of this connective tissue, which we describe by a tendon's force–length curve. We scale these generic curves to represent a specific muscle using five parameters: (1) optimal muscle fiber length, ℓ_o^M; (2) muscle fiber pennation angle at optimal fiber length, ϕ_o; (3) maximum isometric muscle force, F_o^M; (4) maximum muscle contraction velocity, v_{max}^M; and (5) tendon slack length, ℓ_s^T. We begin this chapter by describing each of these five muscle-specific parameters. We will see how each parameter affects muscle force and can be incorporated into a model of muscle–tendon dynamics.

Optimal muscle fiber length, ℓ_o^M

As we saw in Chapter 4, the active force that can be generated by a sarcomere depends on its length (Figure 4.6). The length at which a sarcomere can develop maximum isometric force is called its *optimal length*. Because a muscle fiber is composed of a number (n) of sarcomeres arranged end to end, the fiber also has an optimal length (ℓ_o^M) that is reached when each of its constituent sarcomeres is at its optimal length (ℓ_o^S):

$$\ell_o^M = n\ell_o^S \qquad (5.1)$$

Equation 5.1 assumes that all sarcomeres in series along a muscle fiber are the same length. Muscles lengthen and shorten during movement, which affects the active force they can generate as the thick and thin filaments of their sarcomeres slide past each other. Muscle fibers with longer optimal lengths (i.e., with more sarcomeres in series) have broader active force–length curves and can generate a high fraction of their maximum active force over a wider range of lengths (Figure 5.2, top). Increasing a muscle

108　CHAPTER 5

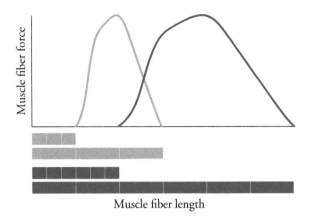

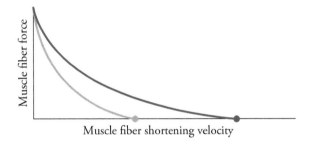
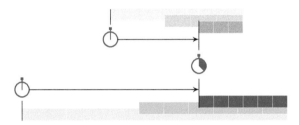

FIGURE 5.2
Muscles with longer optimal fiber lengths have broader active force–length curves (top) and higher maximum shortening velocities (bottom). Note that the schematics of the long fiber (blue) have twice as many sarcomeres as the short fiber (orange), and thus the long fiber can shorten twice the distance in a given time period (bottom right).

fiber's optimal length also increases its maximum shortening velocity (v^M_{max}):

$$v^M_{max} = n v^S_{max} \qquad (5.2)$$

where v^S_{max} is the maximum shortening velocity of a sarcomere. Thus, the force–velocity curve also broadens as the optimal fiber length of a muscle increases (Figure 5.2, bottom).

Biological muscles are comprised of fascicles that have different lengths, which themselves contain fibers that are also of different lengths and that may even terminate within the fascicle. In our model, however, we shall assume that all fibers in a muscle are the same length—an assumption made by many (but not all) models of muscle–tendon dynamics. We further assume that all fibers are straight, parallel, and coplanar. Thus, to characterize the force–length and force–velocity properties of a muscle, we simply magnify the corresponding properties of a muscle fiber which,

MUSCLE ARCHITECTURE AND DYNAMICS   109

in turn, are merely scaled-up versions of the same properties of sarcomeres.

Muscle fiber pennation angle at optimal fiber length, ϕ_o

Muscle typically attaches to bone through tendon. In *parallel-fibered muscles*, such as the sartorius, the fibers are arranged in the direction of the tendon (Figure 5.3). In most other muscles, such as the rectus femoris, the fibers are arranged at an acute angle to the tendon; we refer to these muscles as *pennated*. This word comes from the Latin word for feathered, and indeed the structure of a pennated muscle is reminiscent of a bird feather.

If all muscle fibers attach on one side of the tendon, we say the muscle is *unipennate*; if they attach on both sides, it is *bipennate*. In *multipennate* muscles, the tendon branches and the muscle fiber architecture may be complex. We henceforth assume that all

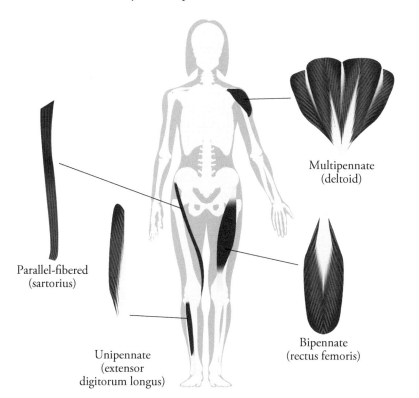

FIGURE 5.3
Examples of muscles with different architectures: parallel-fibered, unipennate, bipennate, and multipennate.

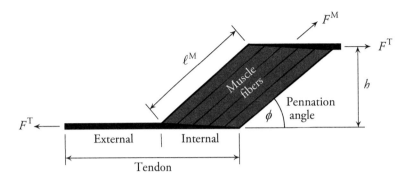

FIGURE 5.4
Simplified geometric representation of muscle fibers and tendon. Muscle fibers are assumed to be straight, parallel, coplanar, of equal length, and attached to tendon at the same pennation angle (ϕ). The height h (and therefore the area) of the parallelogram remains constant as the muscle fibers shorten and lengthen, increasing and decreasing the pennation angle. Adapted from Zajac (1989).

fibers in a given muscle attach to tendon at the same angle, called the *pennation angle* (ϕ), and adopt the model of muscle–tendon geometry shown in Figure 5.4. We thus obtain the following relationship between the force in the muscle fibers (F^M) and the force in the tendon (F^T):

$$F^T = F^M \cos(\phi) \tag{5.3}$$

Now, referring to Figure 5.4, we can explain how biological muscles maintain a constant volume even though the fibers change length. As the fibers in the figure shorten, imagine them shortening in such a way that the parallelogram in Figure 5.4 maintains the same height h. The top of the parallelogram will stay fixed, but the bottom will be pulled to the right, and the parallelogram will become more rectangular. Yet as long as the height remains constant, the area will also remain constant, in accordance with the geometric rule that the area of a parallelogram is the product of its base and height. We have drawn this figure in two dimensions, but the motion is similar in three dimensions and ensures that the volume of the muscle does not change. Stated briefly, pennated muscles maintain their volume not by bulging, but by shearing.

In the process described above, the pennation angle increases and the force transmitted from the fibers to the tendon decreases, until the fibers and tendon are perpendicular and the muscle in the figure becomes rectangular (i.e., $\phi = 90$ degrees). We use the parameter ϕ_o to denote the pennation angle at which the muscle fibers are at their optimal length (i.e., when $\ell^M = \ell^M_o$). The fixed-height approximation that we have described may introduce errors for muscles that bulge appreciably as they contract, but it provides a simple geometric model for studying the functional implications of a muscle's architecture.

Pennation plays a critical role in determining a muscle's force-generating capacity beyond the relationship expressed in Equation 5.3. In general, muscles with a greater pennation angle are able to pack more fibers into a given volume. Consider the analogous situation of installing a hardwood floor in a rectangular room: one could use a relatively small number of long planks that extend the length of the room, or a larger number of shorter planks oriented diagonally. Similarly, compared to a parallel-fibered muscle of the same volume, a pennated muscle will have shorter fibers and, therefore, narrower active force–length and force–velocity curves (Figure 5.2). Of course, the pennated muscle would also contain more fibers, the consequences of which will be explored in the next section.

Maximum isometric muscle force, F_o^M

The third in our series of five muscle parameters is relatively easy to understand but not so easy to measure. This is the *maximum isometric muscle force*. It is defined as the force generated by a muscle when it is maximally activated and held at its optimal fiber length.

In a living person, the maximum isometric muscle force is hard to measure because we cannot isolate one muscle from the others and apply a resisting force only to that muscle. For this reason, we use a proxy called the *physiological cross-sectional area* (PCSA; Figure 5.5). This is the area of a cross-section of the muscle oriented perpendicular to the fibers. Notice that in a pennated muscle, this cross-section is oblique to the longitudinal axis of the muscle. The maximum isometric force can be estimated as follows:

$$F_o^M = \text{PCSA}\, \sigma_o^M \qquad (5.4)$$

where σ_o^M is the muscle's *specific tension* (also known as its *peak isometric stress*), or the maximum muscle force that can be generated per unit area. A typical value for this parameter when modeling healthy muscle is $\sigma_o^M = 0.3$ MPa.

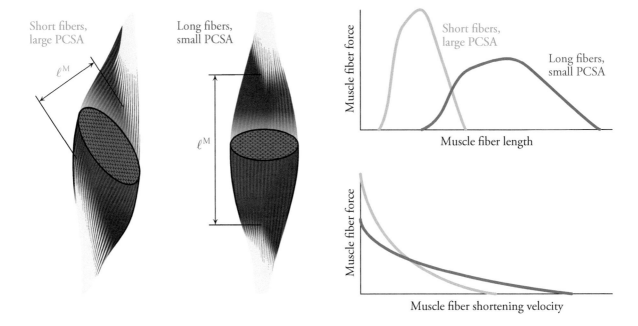

In the previous section, we noted that the fibers in a pennated muscle will be shorter but also more numerous than those in a parallel-fibered muscle of the same volume. Thus, although the force–length and force–velocity curves of the pennated muscle would be narrower, these curves would also be taller because the PCSA (and therefore the maximum isometric force) of the pennated muscle would be greater (Figure 5.5).

The length and contraction velocity of a muscle's fibers are affected by more than just its geometry, as we will see in the next two sections. Nevertheless, we can already observe how a muscle's architecture affects the range of functions it can perform: all else being equal, a pennated muscle will be able to generate greater force than a parallel-fibered muscle of the same volume, but over a smaller range of lengths and at lower contraction velocities.

Maximum muscle contraction velocity, v_{max}^M

Thus far, we have introduced three variables related to the geometric arrangement of a muscle's fibers. To determine the maximum

FIGURE 5.5
The muscles shown on the left have the same volume but different PCSAs, optimal fiber lengths, and pennation angles. The more pennated muscle can generate a greater active force but has shorter fibers; thus, its active force–length and force–velocity curves are taller but narrower. Adapted from Lieber (2010).

contraction velocity of a muscle, we now introduce the idea that muscle fibers come in two types: *fast-twitch* and *slow-twitch*. Fast-twitch fibers have faster rise times and relaxation times in single-twitch experiments (Figure 4.12), and have a higher maximum contraction velocity—roughly 10 optimal-fiber-lengths per second ($10\,\ell_o^M/s$) compared to about $3\,\ell_o^M/s$ for slow-twitch fibers. Mammalian muscles contain both fiber types, but the ratio of fast-twitch to slow-twitch fibers varies depending on a muscle's function. For example, the gastrocnemius muscle has a large fraction of fast-twitch fibers that are recruited during activities that require large forces to be developed quickly, such as sprinting. The neighboring soleus muscle has mostly slow-twitch fibers that are well suited for generating force during long periods of standing, because slow-twitch fibers are resistant to fatigue.

Muscle fibers are distinguished not only by their speed of contraction but also by how they create ATP. Fibers that create ATP aerobically (i.e., using oxygen) are more resistant to fatigue than are fibers that create ATP anaerobically (i.e., without oxygen). Slow-twitch fibers tend to be resistant to fatigue, while fast-twitch fibers fatigue quickly. Humans and some other animals have a third fiber type, so we generally categorize fibers as *Type I* (slow, fatigue-resistant), *Type IIA* (fast, moderately fatigable), or *Type IIB* (very fast, highly fatigable). The fraction of muscle fibers that are fast and only moderately fatigable can be increased with intensive training.

Recall from Chapter 4 that motor units differ in size, and that the central nervous system recruits motor units from smallest to largest, following Henneman's size principle (Figure 4.13). In addition to differing in size, motor units also differ in the types of fibers they contain, with all fibers in a given motor unit being of the same type. The smallest motor units are typically comprised of Type I fibers and are recruited first, followed by the somewhat larger motor units comprising Type IIA fibers. The largest motor units, containing Type IIB fibers, are usually recruited last. Thus, at low activations (as might be observed during quiet standing), primarily slow-twitch, fatigue-resistant motor units are recruited. The maximum muscle contraction velocity is, therefore, likely to

CHAPTER 5

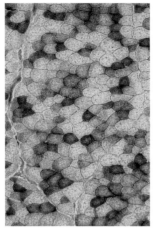

Chicken Swordfish Human

FIGURE 5.6
Some muscles in chicken and fishes comprise primarily fatigue-resistant, slow-twitch fibers (dark meat) while others comprise fatigable, fast-twitch fibers (white meat). In humans and other mammals, fiber types are interspersed within each muscle. Image of human fibers stained to show different fiber types courtesy of Richard Lieber.

decrease at low activations; however, a constant value of $10\,\ell_o^M/s$ is typically assumed in models of muscle–tendon dynamics.

Roughly speaking, fatigable and fatigue-resistant muscle fibers in fowl and fishes are segregated (Figure 5.6). In contrast, fatigable and fatigue-resistant muscle fibers are interspersed in mammalian muscles. This is why a chicken dish can be prepared with either white or dark meat, but a beef dish cannot. The difference in color arises because dark muscle (or meat) is rich in a protein called myoglobin that stores oxygen within the muscle, darkening its color and making it more resistant to fatigue. The dark meat in chicken drumsticks consists of leg muscles that are used for long periods of standing and running (fatigue-resistant, slow-twitch fibers). The white breast meat comprises muscle that is used only for short bouts of flight (fatigable, fast-twitch fibers). In humans, individual muscle fibers are either "dark" or "white," but both fiber types are sprinkled throughout each muscle.

Tendon slack length, ℓ_s^T

So far, none of the parameters we have discussed say anything about tendon. In our Hill-type model, we describe tendons as nonlinear

MUSCLE ARCHITECTURE AND DYNAMICS   115

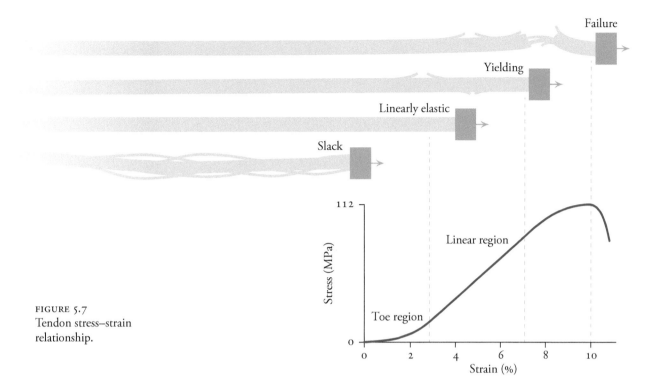

FIGURE 5.7
Tendon stress–strain relationship.

springs. This section introduces the fifth of our five parameters, called the *tendon slack length*, and the fourth of our four generic dimensionless curves, which is called the *tendon force–length curve*. We derive the force–length curve from experimental measurements that relate stress and strain (Figure 5.7).

In this graph, the horizontal axis indicates the tendon's strain, expressed as a percentage increase of its length compared to the length when it is at rest and developing no force. That resting length is what we call the slack length, which we denote by ℓ_s^T. The strain in the tendon (ε^T) at any given instant is defined as follows:

$$\varepsilon^T = \frac{\ell^T - \ell_s^T}{\ell_s^T} \qquad (5.5)$$

where ℓ^T is the tendon's current length and ℓ_s^T is its slack length.

The vertical axis in Figure 5.7 indicates the force generated per unit of cross-sectional area of tendon, called the stress. In a linear spring, the graph of stress against strain is a straight line, whose slope is the *stiffness* of the spring. Because tendon acts as a nonlinear spring, the stress–strain curve is not a straight line, and it has three regions with distinct stiffness characteristics. In the *toe region*, when the tendon is stretched by roughly 0 to 3 percent, it is more compliant ("stretchy"), gradually increasing in stiffness as it lengthens and its constituent collagen fibers uncoil. In the *linear region* of the stress–strain curve, when the tendon is stretched by roughly 3 to 7 percent, the tendon has constant stiffness; in this region it behaves like a linear spring. Finally, above 10 percent strain the tendon begins to experience mechanical failure and there is a high risk of injury. We typically assume that the internal tendon (aponeurosis) and external tendon have the same material properties and strain. There is experimental evidence to support the values of strain given above, but some argue that tendons can experience higher values of strain, up to 15 percent, before they fail.

FIGURE 5.8
Tendon compliance affects muscle force generation. In both cases shown, a parallel-fibered muscle is at its optimal length when the muscle is inactive (top). If the tendon is relatively short and rigid (left), there will be negligible change in the length of the muscle fibers when the muscle is activated. If the tendon is long and compliant (right), the tendon will stretch when the muscle is activated, thereby shortening the muscle fibers and reducing the generated force.

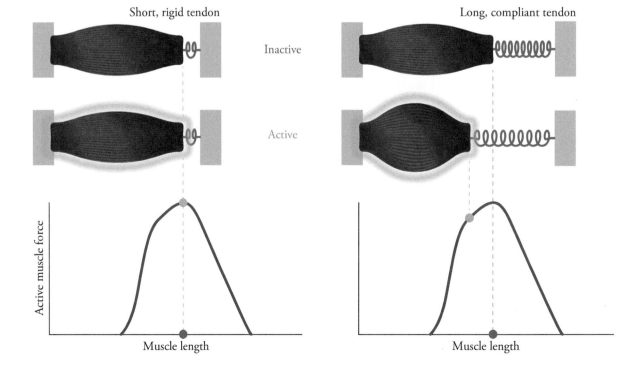

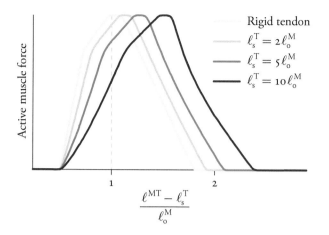

FIGURE 5.9
Effects of tendon compliance on the active force–length curve. Increasing the slack length of the tendon relative to the optimal muscle fiber length (i.e., increasing tendon compliance) broadens the range of lengths over which a muscle–tendon actuator can generate active force.

A tendon can affect the length and therefore the force-generating capacity of the muscle to which it attaches (Figure 5.8). The stretch of a tendon will have little effect if its slack length is short relative to the length of the muscle fibers: even if such a tendon experiences very high strain, its absolute length change (and thus the amount by which the muscle fibers shorten) will be only a small fraction of the optimal length of the muscle fibers. In contrast, if the tendon slack length is long compared to the optimal muscle fiber length, the tendon will stretch substantially when the muscle generates force, resulting in appreciably shorter muscle fibers and a change in the active force that is generated. Figure 5.9 shows that a longer tendon can also increase the range of lengths over which the muscle–tendon unit can generate force. In both of these ways, tendon profoundly affects muscle function. The role of tendon is especially apparent, of course, when a tendon injury renders the muscle practically unusable. As we will see in Chapter 12, tendons also play an important role in the storage and release of energy as they stretch and recoil during running.

Measuring muscle-specific parameters

To summarize, we have defined five muscle-specific parameters, listed in Table 5.1, which capture much of the variability among the muscles of the body. We have listed values for these parameters

Muscle-specific parameter	Symbol	Typical units
Optimal fiber length	ℓ_o^M	cm
Pennation angle at optimal fiber length	ϕ_o	degrees
Maximum isometric force	F_o^M	N
Maximum contraction velocity	v_{max}^M	ℓ_o^M/s
Tendon slack length	ℓ_s^T	cm

TABLE 5.1 Five muscle-specific parameters used by the Hill-type model

for the major muscles of the lower extremity in Table 5.2. For convenience, we often normalize forces, velocities, muscle lengths, and tendon lengths by dividing by F_o^M, v_{max}^M, ℓ_o^M, and ℓ_s^T, respectively, and use a tilde to denote normalized quantities (e.g., $\tilde{F}^M \triangleq F^M/F_o^M$).

Muscle	Maximum isometric force (N)	Optimal fiber length (cm)	Tendon slack length (cm)	Pennation angle at ℓ_o^M (deg)
Adductor brevis	626	10.3	3.5	7
Adductor longus	917	10.8	13.2	8
Adductor magnus				
distal	597	17.7	8.7	11
ischial	597	15.6	21.6	10
middle	597	13.8	4.7	12
proximal	597	10.6	4.0	18
Biceps femoris long head	1313	9.8	32.5	10
Biceps femoris short head	557	11.0	10.6	15
Extensor digitorum longus	603	6.9	36.9	13
Extensor hallucis longus	286	7.5	32.7	11
Flexor digitorum longus	423	4.5	37.9	13
Flexor hallucis longus	908	5.3	35.4	15
Gastrocnemius lateral head	1575	5.9	37.6	12
Gastrocnemius medial head	3116	5.1	39.9	10
Gluteus maximus				
superior	984	14.7	4.9	20
middle	1406	15.7	6.8	21
inferior	948	16.7	7.0	22

TABLE 5.2 Values of muscle-specific parameters for major lower-extremity muscles*

Gluteus medius				
anterior	1093	7.3	5.6	18
middle	765	7.3	6.5	18
posterior	871	7.3	4.5	18
Gluteus minimus				
anterior	374	6.8	1.6	10
middle	395	5.6	2.6	0
posterior	447	3.8	5.1	1
Gracilis	281	22.8	17.2	10
Iliacus	1021	10.7	9.6	16
Peroneus brevis	521	4.5	14.8	12
Peroneus longus	1115	5.1	33.2	14
Piriformis	1030	2.6	11.5	10
Psoas	1427	11.7	10.0	12
Rectus femoris	2192	7.6	44.9	12
Sartorius	249	40.3	12.4	2
Semimembranosus	2201	6.9	34.8	15
Semitendinosus	591	19.3	24.7	14
Soleus	6195	4.4	27.7	22
Tensor fascia latae	411	9.5	45.0	3
Tibialis anterior	1227	6.8	24.1	11
Tibialis posterior	1730	3.8	28.1	13
Vastus intermedius	1697	9.9	20.2	4
Vastus lateralis	5149	9.9	22.1	15
Vastus medialis	2748	9.7	20.0	24

*Maximum isometric forces were derived from muscle volumes measured by Handsfield et al. (2014) in young, healthy subjects. Optimal fiber lengths and pennation angles at optimal fiber length were derived from measurements by Ward et al. (2009) in cadavers. Tendon slack lengths were derived from Rajagopal et al. (2016).

Several techniques have been developed to measure or compute values for muscle and tendon parameters. Architectural parameters have historically been obtained from measurements on human cadavers, whereas dynamic properties such as the shape of the force–velocity relationship have typically been derived from experiments on animal muscles. Care must be taken when obtaining and interpreting measurements from cadavers because tissue properties can change postmortem; for example, muscles become dehydrated, which can change the muscle's mass, volume, shape, and length. Also, cadaveric measurements are typically obtained from the bodies of older donors whose muscles may have atrophied and are, therefore, not representative of the young, healthy subjects who often participate in research studies. Whenever possible, it is generally preferable to calibrate a musculoskeletal model using measurements obtained from the subject being studied, particularly for parameters that are known to vary substantially between individuals (e.g., maximum isometric force).

Muscle fiber lengths and pennation angles can be measured in dissected cadaver specimens or *in vivo* using ultrasonography (Figure 5.10, left). Note, however, that the *optimal* fiber length cannot be determined from ultrasound images because, while the fiber length can be measured, it is not apparent how this length relates to the length at which maximum active force is generated. Measurement of sarcomere lengths provides the information needed to relate measured fiber lengths to optimal fiber lengths, because the sarcomere's force–length curve is known. If muscle fiber length (ℓ^M) and sarcomere length (ℓ^S) are measured, the optimal muscle fiber length (ℓ_o^M) can be computed as follows:

$$\ell_o^M = \ell^M \times \frac{\ell_o^S}{\ell^S} \tag{5.6}$$

where ℓ_o^S is the optimal sarcomere length, estimated to be 2.7 μm in human muscle. There are several techniques to measure sarcomere lengths in whole muscle. One method is laser diffraction, a technique pioneered by Richard Lieber, in which the muscle is surgically exposed and a laser is shone through the muscle to generate a diffraction pattern (Lieber et al., 1984). A less invasive approach developed in my laboratory (Llewellyn et al., 2008) uses needle-sized microendoscopes to image sarcomere lengths *in vivo* (Figure 5.10, right).

Tendon slack lengths cannot be easily determined in cadavers or imaging experiments due to the aponeurosis, which is internal to muscle and difficult to measure. One strategy to determine tendon

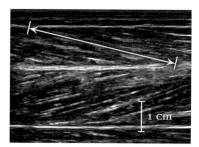 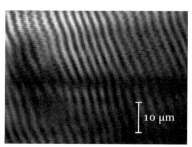

FIGURE 5.10
Imaging provides data needed to calibrate models of muscle–tendon dynamics. Shown at left is an ultrasound image of the fibers of the tibialis anterior muscle from which fiber lengths can be measured. Shown at right is an image of sarcomeres in the same muscle obtained using two-photon microendoscopy, from which sarcomere lengths can be measured. Images courtesy of Glen Lichtwark and Gabriel Sanchez.

slack lengths in models of muscle–tendon dynamics is to measure muscle fiber lengths, sarcomere lengths, and joint positions in cadavers, and then set tendon slack lengths in a model such that the muscle fiber lengths and sarcomere lengths match the cadaveric measurements in the same pose.

As mentioned earlier, we can estimate maximum isometric muscle force from measurements of PCSA and specific tension (Equation 5.4). PCSA can be estimated by measuring muscle mass in cadaver specimens or muscle volume using MRI, the latter being preferable due to the likelihood of atrophy in the muscles of older donor subjects. We can compute PCSA from muscle volume and optimal fiber length as follows:

$$\text{PCSA} = \frac{\text{muscle volume}}{\ell_o^M} \qquad (5.7)$$

Specific tension has been estimated in many experiments on human and animal muscles by computing the ratio of maximum measured muscle tension to a known PCSA or fiber cross-sectional area (see Equation 5.4). Be aware that values for PCSA reported in the literature may, in fact, be strictly geometric cross-sectional areas, or may have already been multiplied by $\cos(\phi)$.

Hill-type model of muscle–tendon dynamics

In this section, we describe a widely used model of muscle–tendon dynamics based on the work of Hill (1938), Wilkie (1956), and Ritchie and Wilkie (1958). A.V. Hill performed the pioneering experiments in the 1930s to characterize the force–velocity property of muscle. The model has since been extended to capture the other salient features of muscle force generation that we have described. The convention is still to call it a "Hill-type model," even though Hill is responsible for only part of it.

The most popular Hill-type model consists of three components: an *active contractile element* and a *passive elastic element* that represent the active and passive force-generating properties of muscle, and a *tendon elastic element* (Figure 5.11A). The symbols for these elements, shown in green, orange, and blue, are borrowed from the engineering literature. We call the combination of these three components a *muscle–tendon actuator*. The model represents muscles and tendons with simple elements; the biological

FIGURE 5.11
Schematic of a typical Hill-type muscle–tendon model (A) and the corresponding generic, dimensionless curves that describe the dynamics of its three components: the force–velocity curve (B), the tendon force–length curve (C), and the active and passive force–length curves (D). The curves shown here are from Millard et al. (2013), which were fit to experimental data and differ from other force–length curves shown in this text.

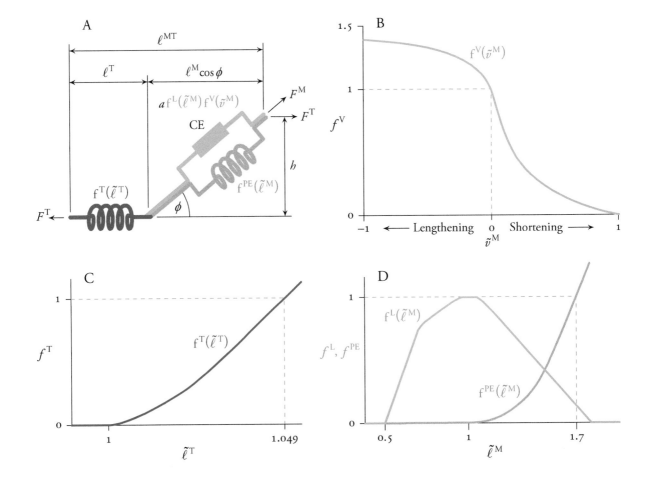

complexities are distilled into the parameters and dimensionless curves that define the dynamics of these elements. Of course, many biological details are omitted in our pursuit of a model that is computationally tractable.

This model is commonly used in studies of whole-body movement because it provides sufficiently accurate estimates of muscle force to study the actions of muscles during walking and running without being computationally burdensome. As we have seen, tendon elasticity can affect muscle fiber dynamics substantially, and the interaction of muscle and tendon is an essential feature of this model. We assume that it is mechanically equivalent to use a single nonlinear spring to represent the effect of the tendons on both ends of the muscle. Note from the schematic in Figure 5.11A that the lengths of the muscle (ℓ^M), tendon (ℓ^T), and muscle–tendon actuator (ℓ^{MT}) are related as follows:

$$\ell^{MT} = \ell^M \cos(\phi) + \ell^T \qquad (5.8)$$

Dimensionless curves

The force-generating properties of the three components in the Hill-type model are described by four generic, time-invariant curves, listed in Table 5.3 and shown in Figure 5.11B–D. These curves become dimensionless once we normalize them by muscle-specific parameters.

The generic curves are derived from experimental measurements and reflect the biological properties of muscle and tendon. In particular, the active force–length curve is a scaled version of the relationship between the length of a sarcomere and the force it generates when activated (Figure 4.6); the hyperbolic force–velocity curve reflects the dynamics of cross-bridge cycling (Figure 4.9);

TABLE 5.3
Four dimensionless curves used by the Hill-type model

Model element	Dimensionless curve	Symbol
Active contractile	Active force–length	$f^L(\tilde{\ell}^M)$
	Force–velocity	$f^V(\tilde{v}^M)$
Passive elastic	Passive force–length	$f^{PE}(\tilde{\ell}^M)$
Tendon elastic	Tendon force–length	$f^T(\tilde{\ell}^T)$

the passive force–length curve represents the force generated when a muscle is stretched beyond its resting length, regardless of its activation (Figure 4.7); and the tendon force–length curve expresses the nonlinear relationship between a tendon's strain and its stress (Figure 5.7). As indicated by the normalized variables appearing in Figure 5.11, the four generic, dimensionless curves are scaled by the tendon's slack length (ℓ_s^T) and the muscle's maximum isometric force (F_o^M), optimal fiber length (ℓ_o^M), and maximum contraction velocity (v_{max}^M). Tendon force scales with the muscle's maximum isometric force because, in general, stronger muscles have stronger tendons to avoid tendon failure.

The importance of scaling and using dimensionless curves could be discussed at great length, but the main point is that scaling allows us to isolate the part of the model that is unique to each muscle and individual. What remains is the part that is generic, shared by all healthy muscles.

Computing muscle force with a rigid tendon

We use the Hill-type model shown in Figure 5.11 to generate muscle-driven simulations of movement. In this section, we present the mathematical details that make up the "contraction dynamics" block in Figure 4.15. Specifically, we describe a strategy to compute muscle force if we know muscle activation and the length and velocity of the muscle–tendon actuator. We will assume here that muscle activation has already been computed by the activation dynamics model, thus $a(t)$ is known. We also assume that we know the length and velocity of the entire muscle–tendon actuator ($\ell^{MT}(t)$ and $v^{MT}(t)$). However, we must determine the muscle and tendon lengths. Once these lengths are known, we can use the muscle-specific parameters with the dimensionless curves shown in Figure 5.11 to compute the forces being generated by the muscle and transmitted through the tendon. (Note that we often omit the "(t)," indicating functional dependence on time, for simplicity of notation but show it explicitly in this section and the next to distinguish time-varying quantities from constants. Also note that we have defined muscle fiber velocity as shortening velocity, in keeping with the convention used in the muscle physiology literature and in Chapter 4. Consequently, $\dot{\ell}^M(t) = -v^M(t)$ in the equations below.)

Let's begin with a simpler problem and assume that the tendon is rigid; we will then generalize our strategy in the next section to account for elastic tendon. The rigid-tendon assumption replaces the tendon spring shown in Figure 5.11A with a rigid rod (i.e., ℓ^T is constant). This assumption is reasonable if the tendon is short relative to the muscle to which it attaches, in which case the tendon will not stretch appreciably beyond its slack length even when subjected to high tensile forces (recall Figure 5.8). We are then left with two unknowns in Equation 5.8, and can relate muscle length ($\ell^M(t)$) to muscle fiber pennation angle ($\phi(t)$) as follows:

$$\ell^M(t) = \frac{\ell^{MT}(t) - \ell^T}{\cos(\phi(t))} \qquad (5.9)$$

We have one equation but two unknowns. To make headway, recall that the fixed-height pennation model relates muscle length and pennation angle to the height of the parallelogram (h) shown in Figure 5.4, which remains constant:

$$h = \ell^M(t) \sin(\phi(t)) \qquad (5.10)$$

We can solve for height h by substituting into Equation 5.10 the optimal fiber length (ℓ_o^M) and the pennation angle at optimal fiber length (ϕ_o):

$$h = \ell_o^M \sin(\phi_o) \qquad (5.11)$$

We can now solve Equations 5.9 and 5.10 for muscle length ($\ell^M(t)$) and pennation angle ($\phi(t)$), where parameter h is given by Equation 5.11. We know this system is solvable because we have as many independent equations as unknowns. Substituting Equation 5.9 into Equation 5.10 produces an expression for the pennation angle in terms of known quantities:

$$\phi(t) = \tan^{-1}\left(\frac{h}{\ell^{MT}(t) - \ell^T}\right) \qquad (5.12)$$

Finally, we can solve for muscle length using Equation 5.9.

Muscle force ($F^M(t)$) can be computed from the activation ($a(t)$), normalized muscle length ($\tilde{\ell}^M(t)$), and normalized muscle velocity ($\tilde{v}^M(t)$):

$$F^M(t) = F_o^M \left[a(t)\, f^L\!\left(\tilde{\ell}^M(t)\right) f^V\!\left(\tilde{v}^M(t)\right) + f^{PE}\!\left(\tilde{\ell}^M(t)\right) \right] \quad (5.13)$$

Here we have summed the active and passive components of the muscle force because they act in parallel. To compute the muscle velocity ($v^M(t)$), we note that velocity is the time derivative of length and differentiate Equation 5.8:

$$v^{MT}(t) = -v^M(t)\cos(\phi(t)) - \ell^M(t)\,\dot\phi(t)\sin(\phi(t)) + v^T(t) \quad (5.14)$$

Recall that we have assumed $v^{MT}(t)$ is known, and the rigid-tendon assumption tells us that $v^T(t) = 0$. Thus, there are only two unknowns in Equation 5.14: muscle velocity ($v^M(t)$) and pennation angular velocity ($\dot\phi(t)$). Two unknowns but only one equation. Fortunately, we can also differentiate Equation 5.10 to obtain a second equation relating these variables:

$$\dot\phi(t) = \frac{v^M(t)}{\ell^M(t)} \tan(\phi(t)) \quad (5.15)$$

We now substitute Equation 5.15 into Equation 5.14 to solve for muscle velocity, and finally use Equation 5.13 to compute muscle force.

Computing muscle force with a compliant tendon

The situation is more complicated when the tendon is compliant rather than rigid. In this case, the tendon can change length, and thus $\ell^T(t)$ is an unknown in Equation 5.9. One strategy for accommodating this additional unknown is to define muscle length $\ell^M(t)$ as a state variable—that is, a quantity that will be numerically integrated from one instant to the next over the course of a simulation. This process works as follows. We provide an initial value for muscle length at time zero ($\ell^M(0)$), an equation that describes the rate at which muscle length is changing (i.e., its velocity, $v^M(t)$), and a numerical integrator for computing the

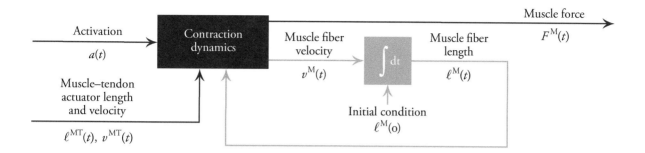

FIGURE 5.12
Muscle fiber length ($\ell^M(t)$) is integrated forward in time to compute contraction dynamics with a compliant tendon.

muscle length a short time into the future from its current length and velocity. The "contraction dynamics" block in Figure 4.15 will then have the form shown in Figure 5.12.

Using a simple numerical integrator, we might compute the future muscle length as follows:

$$\ell^M(t + \Delta t) = \ell^M(t) - \Delta t \, v^M(t) \quad (5.16)$$

where Δt is the amount of time we advance into the future with each time step. (Recall that $\dot{\ell}^M(t) = -v^M(t)$.) By repeating this process over and over, we eventually obtain the value of muscle length over the entire time interval of interest. Use of the integration strategy shown in Equation 5.16 (known as Euler's method) generally requires that we take small time steps to obtain accurate answers.

We can now derive an expression for $v^M(t)$ by assuming that the muscle and tendon are massless and frictionless, in which case there is no net horizontal force at the point where they meet. Thus, the muscle and tendon are in equilibrium:

$$F^T(t) = F^M(t) \cos(\phi(t)) \quad (5.17)$$

(Because the tendon is constrained to remain horizontal in Figure 5.11A, we can ignore the vertical component of the force generated by the muscle, $F^M(t) \sin(\phi(t))$.) The force generated by the tendon ($F^T(t)$) can be computed from normalized tendon length ($\tilde{\ell}^T(t)$) using the tendon force–length curve:

$$F^T(t) = F_o^M \left[f^T\left(\tilde{\ell}^T(t)\right) \right] \quad (5.18)$$

We obtain an expression for muscle velocity by substituting expressions for the muscle and tendon forces (Equations 5.13 and 5.18) into the equilibrium equation (Equation 5.17) and solving for $\tilde{v}^M(t)$:

$$\tilde{v}^M(t) = f^V_{inv}\left(\frac{f^T(\tilde{\ell}^T(t))/\cos(\phi(t)) - f^{PE}(\tilde{\ell}^M(t))}{a(t)\,f^L(\tilde{\ell}^M(t))}\right) \quad (5.19)$$

where f^V_{inv} is the inverse of the force–velocity curve (i.e., it describes muscle velocity as a function of force). Note that Equation 5.19 has four numerical singularities that must be avoided during a simulation: as the pennation angle approaches 90 degrees, as activation approaches zero, at fiber lengths where the active force–length curve approaches zero, and where the force–velocity curve is not invertible. These singularities can be avoided by introducing constraints on the corresponding variables, such as by preventing pennation from exceeding some upper limit less than 90 degrees.

The approach described above is used to estimate muscle and tendon forces in nearly all muscle-driven simulations of movement. The approach is valuable because it allows us to account for each muscle's activation, fiber length, and fiber velocity in estimating the force it generates. However, other approaches are also valuable, which we describe below.

Other models of muscle force generation

We have focused on the Hill-type model because it is widely used in simulations of movement, including those presented throughout this book. However, several phenomena are not captured by the Hill-type model. For example, the model ignores *short-range stiffness*, which describes a muscle's velocity-independent resistance to small, rapid perturbations in its length; *force enhancement*, or the increase in a muscle's maximum isometric force immediately following stretch; and dependence on temperature and fatigue. Furthermore, although the Hill-type model does a good job of representing changes in muscle force with changes in length and velocity when the muscle is maximally active, it is less accurate for submaximal activation (Millard et al., 2013). These limitations must be considered when using a Hill-type model in scientific investigations.

Other muscle models have been developed, some simpler and some more complex. A simple strategy for modeling muscle forces is to represent the total moment generated by all muscles crossing a joint with a single torque actuator (a motor with physiologically inspired properties). The actuator's maximum torque can be expressed as a function of the joint angle, as determined experimentally, but individual muscle forces cannot be computed. As we will see, this model is sufficiently simple to use in large-scale optimization problems, yet still produces simulations that closely resemble human movement.

There are also muscle models that provide more detailed information when these details are necessary to answer a particular question. A well-known example is based on the work of Andrew Huxley (1957), in which the sliding-filament mechanism of muscle contraction is modeled explicitly. This *Huxley-type model* describes force generation using partial differential equations of the following form:

$$\frac{\partial n(x,t)}{\partial t} - v(t)\frac{\partial n(x,t)}{\partial x} = f(x) - \big(f(x) + g(x)\big)n(x,t) \quad (5.20)$$

where $n(x,t)$ is a probability density function describing the proportion of actin and myosin that are engaged in cross-bridge cycling; x and t are the independent variables of distance and time; $v(t)$ is the shortening speed of a half-sarcomere; and the functions $f(x)$ and $g(x)$ represent, respectively, the rates at which cross-bridges are formed and broken. A major challenge in using this model is to determine appropriate values for the many parameters in the model, since little is known about how these quantities vary between species and between muscles. Nevertheless, this model captures some of the effects noted above that are not captured by the Hill-type model. The cross-bridge model is usually used to study the dynamics of a single muscle.

Finite-element models provide yet another alternative. In a finite-element model, the muscle is divided into many small elements (on the order of thousands), each of which is governed by a set of equations that describe the behavior of the tissue. Finite-element models are the state of the art for engineering applications, such as the design of aircraft, and without question they would provide greater accuracy than the simple equations we have derived in this chapter.

But greater power comes at a cost: finite-element models can take hundreds of times longer to solve than Hill-type models. Despite this computational burden, finite-element models can be used to good effect to model muscles whose fibers have different lengths, strain nonuniformly, and transmit tension laterally (Figure 5.13). Ultimately, you must find the right balance between the amount of detail provided by a model and its analytical tractability for the application at hand.

Since many muscles are involved in the coordination and control of movement, simulations with large numbers of muscles are critical for studying activities like walking and running. The models discussed in this chapter allow us to simulate dozens of muscles at a time using only modest computational resources. We will flex our computational muscles in Chapters 10–12 to generate muscle-driven simulations of movement.

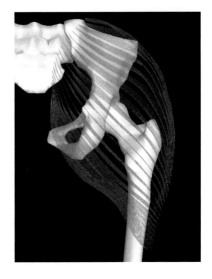

FIGURE 5.13
A model of the gluteus maximus from Blemker and Delp (2005). Finite-element models provide detailed characterizations of muscle architecture and allow calculation of internal muscle and tendon strains.

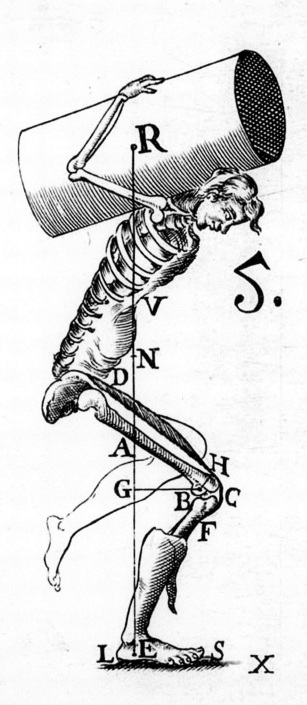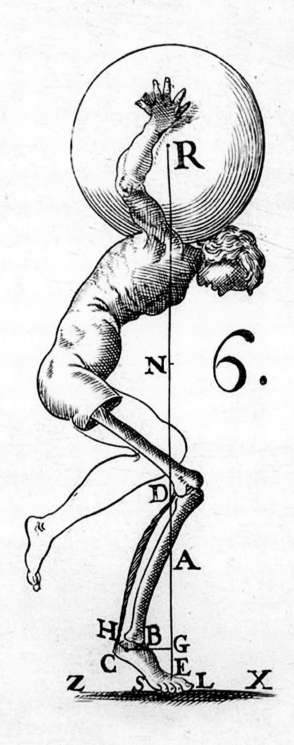

6 Musculoskeletal Geometry

Give me a place to stand and with a lever I will move the whole world.

—Archimedes

WHEN I WAS IN GRADUATE SCHOOL, one of my mentors was Eugene Bleck, the chief of orthopedic surgery at Stanford Hospital and the person who literally wrote the book on orthopedic management of cerebral palsy. One of the most common causes of disability in children, cerebral palsy affects a person's ability to walk and maintain balance. Individuals with cerebral palsy often develop distinctive gaits, like crouch gait, which are inefficient and exhausting.

In many cases, surgery to lengthen the hamstring muscles can improve crouch gait. But not always! When I asked Dr. Bleck how he decided whether to lengthen the hamstrings of a particular patient, he told me that he lengthens the muscles when they are "too tight," which he assessed by seeing how far he could stretch the muscles with the children lying on his exam table. As I traveled around to different hospitals, I found that their criteria for performing the surgery were inconsistent. I thought there must be a better way to determine which children would benefit and which ones would not. Was the problem simply that the hamstrings were too short, or were there other issues that would not be resolved by making the muscle longer?

The computer models we began developing in Chapters 4 and 5 are well suited to answer such questions. Before performing an irreversible surgery on a person, perhaps we could simulate the surgery on a computational model. But we need to add something to these models first: the way that muscles interact with the skeleton to produce motion of the whole body.

We start with the fact that muscles produce movement by applying forces to bones. However, there is a subtlety here. If, for example, your biceps attached exactly at your elbow joint, you would never be able to move your forearm. Instead, muscles attach to bones at a distance from anatomical joints so that they have sufficient leverage to move the limbs (if not the whole world). This leverage is called the muscle's mechanical advantage or *moment arm*. The moment arm can be increased by increasing the distance between the muscle attachment site on the bone and the joint's axis of rotation. The effect is similar to placing a door handle far from the hinge to make it easy to open the door. The mechanical advantage of a muscle is determined by bone geometry, body pose, and the path of the muscle's line of action. We refer to these features collectively as *musculoskeletal geometry*.

As we will see in this chapter, a muscle's mechanical advantage is intimately related to its functions in the body. Thus, musculoskeletal geometry is studied by scientists and clinicians from a variety of disciplines. Paleontologists study the shapes of fossilized bones from extinct animals to determine where muscles attached and to make inferences about how these animals might have moved. Biomechanists study the musculoskeletal geometry of world-class athletes to determine the features that make elite performance possible. And surgeons can apply musculoskeletal models to improve the lives of people with cerebral palsy.

We will begin our journey with a simple example, which will nevertheless teach us a lot about how these models work. The questions are: How much effort does it take to stand still? And which is harder: standing still with your weight over the front of your feet, or standing still with your weight centered on the middle of your feet?

Muscle mechanical advantage

To answer these questions, we consider a two-dimensional, static analysis of the force required by a single ankle plantarflexor muscle

FIGURE 6.1
Free-body diagrams (left) and model (right) used to estimate plantarflexor muscle force with a person standing in an upright posture (F_1^M) and when leaning forward (F_2^M).

to hold the body upright during quiet standing. We model the body as a single inverted pendulum, assume the entire weight of the body is supported by a single leg, and compare two scenarios. In the first scenario, the person stands with her weight evenly distributed over her foot, which positions the center of pressure halfway between her heel and toes (Figure 6.1, left). In the second scenario, she leans forward, which moves the center of pressure toward her toes.

We begin our analysis by drawing a free-body diagram of the foot in each scenario, isolating it from the rest of the body. We represent the foot as a triangle and assume that the ankle joint is located at the upper vertex A of the triangle. According to Archimedes' law of the lever, the fact that we can stand upright without falling either forward or backward means that the sum of the moments about the ankle joint is zero:

$$\sum M_A = F_i^{GRF} \ell_i - w_{foot} c - F_i^M r_i = 0 \quad (6.1)$$

The index i refers to the two scenarios, 1 and 2, in which the person is standing with her weight centered over her foot and leaning forward, respectively. Rearranging this equation allows us to calculate the force that must be generated by the muscle:

$$F_i^M = \frac{F_i^{GRF} \ell_i - w_{foot} c}{r_i} \quad (6.2)$$

This equation shows that the force in the muscle (F_i^M) is influenced by the magnitude of the ground reaction force (F_i^{GRF}) and the

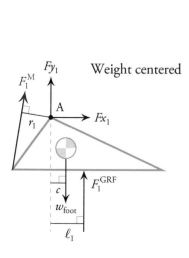
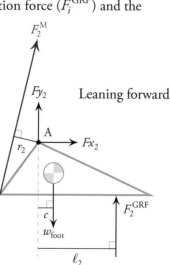
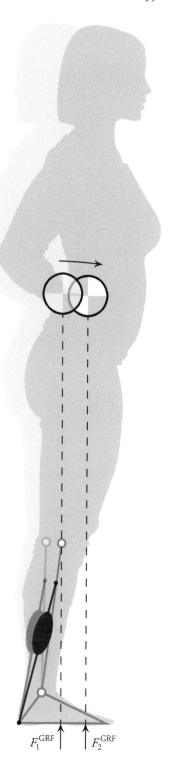

weight of the foot (w_{foot}), as well as the geometry of the system: the location of the center of pressure (ℓ_i), the location of the foot's center of mass (c), and, notably, the perpendicular distance from the center of the ankle to the muscle force vector (r_i). The distance r_i defines the muscle moment arm. A muscle with a small moment arm must apply greater force to generate the same joint moment as a muscle with a large moment arm. We say that the muscle with the larger moment arm has a greater mechanical advantage.

Realistic quantities for each variable in each scenario are shown in Table 6.1. In the first scenario, the force in the muscle is roughly equal to the magnitude of the ground reaction force (which is equal to body weight because the person is in static equilibrium). In the second scenario, simply moving the center of pressure forward by 3 centimeters results in a doubling of the muscle force.

TABLE 6.1
Calculation of muscle forces during standing in two scenarios

Scenario	F_i^{GRF}	w_{foot}	ℓ_i	c	r_i	F_i^{M}
Weight centered ($i=1$)	600 N	7.8 N	5 cm	1 cm	5 cm	598 N
Leaning forward ($i=2$)	600 N	7.8 N	8 cm	1 cm	4 cm	1198 N

Note that we balanced moments about the ankle to compute muscle force F_i^{M} (Equation 6.2). If we instead balance forces in the horizontal and vertical directions, we can compute the two components of the joint reaction force (Fx_i and Fy_i). We find that the vertical force (Fy_i) is about 2–3 times body weight even during quiet standing. Large internal joint contact forces exist because the moment arms of muscles are generally small; thus, large muscle forces can be required even when engaging in ostensibly effortless tasks like standing still or walking slowly.

The above example applies basic concepts of mechanics to a simple two-dimensional static analysis involving only one muscle. In the next section, we develop a more formal engineering definition of a moment arm that applies more broadly to three-dimensional problems.

Definition of a muscle moment arm

In our analysis of musculoskeletal geometry, we assume that muscle–tendon units can be modeled as frictionless, extensible strings that attach to bones and wrap around anatomical structures. Although biological muscles are composed of thousands of fibers, each with a unique three-dimensional path, we often represent a muscle as a single tensile force acting along the path from one point of attachment to another. This path determines the points at which the muscle force is applied as well as the direction of application. The moment generated by a muscle ($\underline{M}$) can be computed as follows (see Figure 6.2):

$$\underline{M} = \underline{r} \times \underline{F} \qquad (6.3)$$

where "×" denotes the vector cross product. Point O indicates the joint's center of rotation, vector $\underline{F}$ represents the magnitude and orientation of the muscle force applied at point P_1, and $\underline{r}$ is a vector from O to any point on the muscle's line of action.

One appealing property of the straight-line muscle model, illustrated in Figure 6.2, is the fact that $\underline{r}$ may be drawn from O to P_1, or to P_2, or to any other point along the muscle's line of action.

FIGURE 6.2
Definition of moment arm r associated with the generation of a moment about point O.

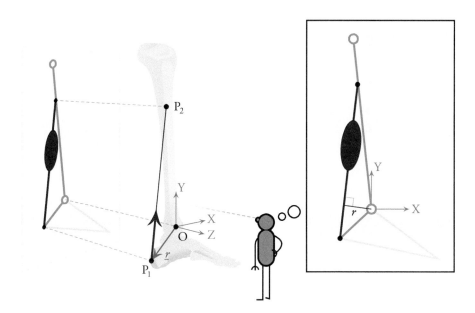

In all cases, the cross product in Equation 6.3 will come out the same. The moment generated by a muscle ($\underline{M}$) is a vector, but it is often more convenient to obtain a scalar quantity. In this example, we project Equation 6.3 onto the Z axis:

$$M_z = (\underline{r} \times \underline{F}) \cdot \hat{z} \qquad (6.4)$$

where "·" denotes the vector dot product and $\hat{z}$ is a unit vector along the Z axis, the axis of rotation associated with moment M_z. We define the moment arm of the muscle (r) by normalizing M_z by the magnitude of the muscle force ($\|\underline{F}\|$):

$$r \triangleq \frac{\underline{r} \times \underline{F}}{\|\underline{F}\|} \cdot \hat{z} \qquad (6.5)$$

Remember that r represents the perpendicular distance from the muscle's line of action to the joint's center of rotation. In Figure 6.3, you can see how to estimate this distance from a magnetic resonance imaging (MRI) scan of the ankle. Here we have to assume that the axis of rotation of the ankle, $\hat{z}$, is perpendicular to the plane of the image. If the ankle is oblique to the scanner, then our estimate of r will not be accurate. For that reason, and because we do not always have an MRI scan available, we will present another strategy for estimating r in the next section.

The perpendicular distance between the muscle's line of action and the axis of rotation changes when the joint angle changes (Figure 6.4). Consequently, a muscle's moment arm varies as a function of the joint angle. For many limb muscles, this function reaches a peak value near the middle of the range of motion; however, the relationship varies widely across muscles and must be measured if we are to accurately characterize the moment arm.

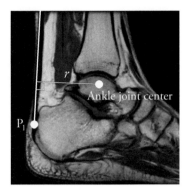

FIGURE 6.3
Estimating muscle moment arm from a magnetic resonance image. MRI courtesy of Alex Frietas.

Tendon-excursion definition of a moment arm

We now introduce a second definition of the moment arm: the tendon-excursion definition. Given certain assumptions, this definition is mathematically equivalent to the anatomical definition described in connection with Figure 6.3; however, the tendon-excursion definition provides valuable conceptual insights.

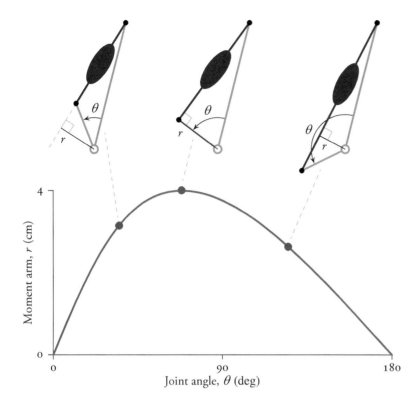

FIGURE 6.4
Demonstration of how a muscle's moment arm (r) depends on the angle of the joint that it spans (θ).

Consider the system shown in Figure 6.4, where the muscle path is represented by a line segment. This representation makes three assumptions: (1) the path of the muscle is independent of muscle force (which may not be the case for muscles that bulge appreciably as they contract), (2) the path of the muscle can be defined by a series of line segments (which may not be the case for large muscles with broad attachments or complicated path geometry), and (3) the muscle–tendon unit slides without friction over neighboring structures (which may not be the case for muscles that are attached to neighboring muscles with connective tissue).

As described by An et al. (1984), the tendon-excursion definition arises as a result of applying the principle of virtual work to a muscle–joint system. The principle of virtual work states that the total work done by all forces and moments acting on a system in static equilibrium is zero for a set of infinitesimally small "virtual displacements" (imagined displacements occurring with time held fixed). The virtual work done by a muscle (δw) is defined as the product of the force generated by the muscle (F^M) and a virtual translational displacement of the muscle–tendon unit ($\delta \ell^{MT}$):

$$\delta w = F^M \, \delta \ell^{MT} \tag{6.6}$$

For the total work done by all forces and moments acting on this system to be zero, the virtual work done by the muscle must be balanced by the virtual work done by an external moment, defined as follows:

$$\delta w = M \, \delta \theta \tag{6.7}$$

where $\delta \theta$ is the virtual rotational displacement of the joint. Equating the right-hand sides of Equations 6.6 and 6.7 balances the virtual work done by the muscles with the virtual work done by external moments; substituting Equation 6.3 then leads to the following relationship:

$$\begin{aligned} M \, \delta \theta &= F^M \, \delta \ell^{MT} \\ r \, F^M \, \delta \theta &= F^M \, \delta \ell^{MT} \\ r &= \frac{\delta \ell^{MT}}{\delta \theta} \end{aligned} \tag{6.8}$$

Finally, we take the limit as the virtual displacement goes to zero, whereupon this equation takes the form of a partial derivative:

$$r = \frac{\partial \ell^{MT}}{\partial \theta} \tag{6.9}$$

Equation 6.9 enables us to measure moment arms experimentally using the tendon-displacement method. In this method, the length of the muscle–tendon unit is monitored with a position transducer while the joint moves through a range of motion; the joint angle is recorded simultaneously. Thus, we obtain samples of how the muscle–tendon length varies with joint angle (Figure 6.5, left). To determine the relationship between moment arm and joint angle from these data, we can simply compute the derivative of muscle–tendon length with respect to joint angle, in radians, as described by

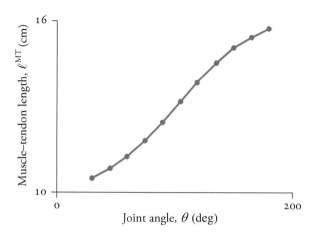
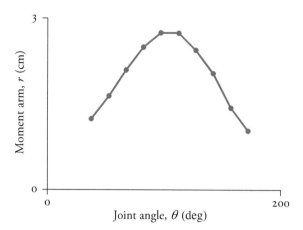

FIGURE 6.5
Tendon-excursion data and the corresponding moment-arm data, calculated using numerical derivatives. These curves correspond to the data shown in Table 6.2.

Equation 6.9. The data shown in Table 6.2 and the corresponding plot in Figure 6.5 demonstrate how the method of central differences can be used to calculate muscle moment arm as a function of joint angle. Notice that if we wanted to determine this relationship using the "anatomical" method shown in Figure 6.3, we would need to take many MRI scans of the ankle with different amounts of flexion.

We and others have measured tendon excursions of many muscles in the arms and legs, and have used these data to estimate the moment arms of these muscles. For example, Wendy Murray

θ (deg)	ℓ^{MT} (cm)	$\Delta\theta$ (rad)	$\Delta\ell^{MT}$ (cm)	θ at midpoint (deg)	Moment arm (cm)
30	10.50				
45	10.82	$\pi/12$	0.32	37.5	1.22
60	11.25	$\pi/12$	0.43	52.5	1.64
75	11.80	$\pi/12$	0.55	67.5	2.10
90	12.45	$\pi/12$	0.65	82.5	2.48
105	13.17	$\pi/12$	0.72	97.5	2.75
120	13.89	$\pi/12$	0.72	112.5	2.75
135	14.53	$\pi/12$	0.64	127.5	2.44
150	15.06	$\pi/12$	0.53	142.5	2.02
165	15.44	$\pi/12$	0.38	157.5	1.45
180	15.71	$\pi/12$	0.27	172.5	1.03

TABLE 6.2
Calculation of moment arm from tendon-excursion data

measured tendon excursions and moment arms for the major muscles that cross the elbow (Figure 6.6). These data are useful for testing the accuracy with which computer models of the musculoskeletal system represent the geometry of humans and for determining the rules that describe how musculoskeletal geometry differs between individuals.

FIGURE 6.6
Tendon excursions and moment arms for major muscles crossing the elbow. Data from Murray et al. (1995).

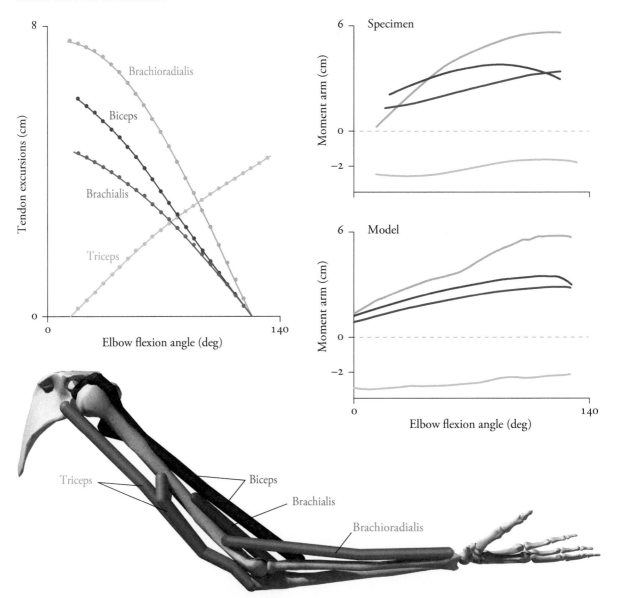

Muscle moment arms affect muscle lengths and velocities

The tendon-excursion definition of a moment arm demonstrates that, over a given range of joint motion, a muscle–tendon unit will undergo a greater change in length if its moment arm is larger. Because a muscle's force-generating capacity depends on its length (due to the force–length relationship), a muscle with a large moment arm will operate over a larger range of the force–length curve than a muscle with a small moment arm (assuming the muscles have equal optimal fiber lengths; Figure 6.7).

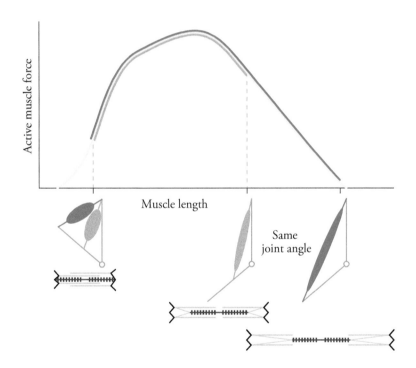

FIGURE 6.7
Two muscles with the same optimal fiber length but different moment arms change length by different amounts and, therefore, traverse different ranges of the active force–length curve for the same range of joint angle.

A muscle's moment arm also affects the velocity of the muscle–tendon unit (v^{MT}) for a given joint angular velocity ($\partial \theta / \partial t$). The relationship between v^{MT} and $\partial \theta / \partial t$ can be examined mathematically by first multiplying Equation 6.9 by the joint angular velocity:

$$\left(\frac{\partial \ell^{MT}}{\partial \theta}\right)\left(\frac{\partial \theta}{\partial t}\right) = r \frac{\partial \theta}{\partial t} \qquad (6.10)$$

Simplifying Equation 6.10, we obtain the following relationship:

$$\frac{\partial \ell^{\mathrm{MT}}}{\partial t} = v^{\mathrm{MT}} = r\frac{\partial \theta}{\partial t} \qquad (6.11)$$

Equation 6.11 demonstrates that, for a given joint angular velocity, a muscle with a large moment arm will experience higher velocities than one with a small moment arm. This relationship has important functional implications because muscle force depends on muscle velocity, as described by the force–velocity curve (Figure 6.8).

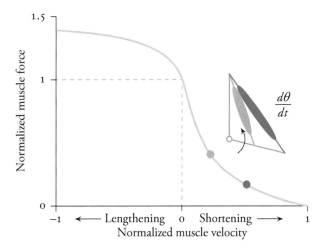

FIGURE 6.8
A muscle with a larger moment arm (blue) will shorten at a higher velocity than a muscle with a smaller moment arm (orange). The force-generating capacity of the muscle with the larger moment arm will be lower due to the force–velocity relationship.

The effect of muscle moment arm on muscle velocity suggests that it might be beneficial for muscles to have small moment arms and long fibers if they actuate joints that undergo high angular velocities, to reduce the velocity at which the fibers shorten. Indeed, Sabrina Lee and Steve Piazza (2009) found that elite sprinters have Achilles tendon moment arms that are 25 percent smaller and ankle muscle fibers that are 11 percent longer than those of size-matched non-sprinters. Longer muscle fibers and smaller moment arms allow elite sprinters to maintain plantarflexor fiber lengths that are closer to their optimal lengths and to reduce fiber shortening velocities as the ankle plantarflexes rapidly during the push-off phase of running. These morphological features increase the force-generating capacity of the plantarflexors, which enables them to generate the large forces necessary for high-speed running.

In addition to comparing the same muscle in different individuals, it is instructive to compare the capabilities of different muscles in the same individual and how they are affected by moment arms. Increasing the moment arm of a muscle will increase the magnitude of the maximum moment it can generate but will also increase the amount (and rate) by which it changes length for a given movement. Larger changes in length can result in decreased muscle force-generation capacity at the extremes of the joint's range of motion as the muscle fibers reach the extremes of their force–length relationship. These two phenomena create a trade-off between maximum moment and maximum joint excursion. Different muscles will operate at different points in the range of possible moment arms, making each uniquely suited to perform a particular set of functions and, collectively, enabling the remarkable versatility of the human body.

Moment arms of multi-joint muscles

Thus far, we have limited our analysis to muscles that cross only one joint, yet many muscles cross more than one joint and have moment arms with respect to each. For example, the semimembranosis, one of the hamstring muscles, crosses behind the hip and knee joints and thus has two moment arms in a planar model (Figure 6.9). In fact, even if a muscle spans only one joint, it will have one moment arm for each degree of freedom of the joint. For example, the gluteus maximus, which crosses only the hip, has moment arms for hip extension, hip abduction, and hip rotation. The moment arm for each degree of freedom can be calculated with Equation 6.4.

The amount by which a muscle changes length as the body moves depends on the displacements of the joints spanned by the muscle and the moment arms of the muscle with respect to each joint angle. The velocity of a muscle–tendon unit depends on the angular velocities of the joints spanned by the muscle and, again, on the moment arms of the muscle with respect to each joint angle. We can generalize Equation 6.11 to calculate the

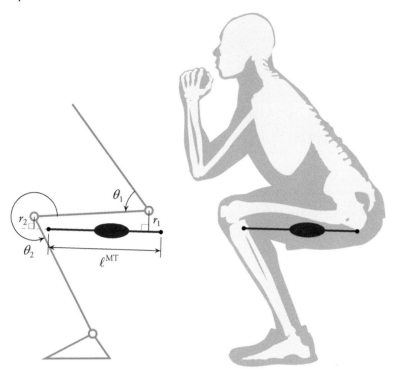

FIGURE 6.9
The hamstrings cross posterior to the hip and knee. The muscle shown here has a hip extension moment arm (r_1) and a knee flexion moment arm (r_2). Its length (ℓ^{MT}) and velocity depend on the joint angles (θ_1 and θ_2) and angular velocities.

velocity of a muscle–tendon unit (v^{MT}) that crosses n joints as follows:

$$v^{MT} = \sum_{i=1}^{n} r_i \frac{d\theta_i}{dt} \qquad (6.12)$$

where r_i is the moment arm corresponding to the ith degree of freedom (θ_i).

As we mentioned at the beginning of this chapter, calculations of muscle–tendon lengths and velocities can guide treatment of movement disorders. This may be helpful for individuals with cerebral palsy who walk with a crouch gait, in which the knee is flexed more than normal (Figure 6.10, left). This gait abnormality is often attributed to "short" or "spastic" hamstring muscles that restrict knee extension and is often treated by surgically lengthening the hamstrings. Allison Arnold and I worked with collaborators from several children's hospitals to examine whether the hamstrings are shorter than normal in subjects with crouch gait, which would suggest that they might be good candidates for hamstring lengthening surgery. We examined 152 children during crouch gait, measuring their joint angles and stride durations, and then used a computer model to calculate the muscle–tendon lengths and velocities of the hamstrings over the gait cycle. We compared our results to the same lengths and velocities calculated for 45 unimpaired individuals. All lengths and velocities were normalized to account for the different sizes of the participants. The resulting plots enabled us to distinguish between the subjects whose hamstrings were operating at shorter peak lengths than normal and those whose hamstrings were operating at slower peak velocities

FIGURE 6.10
Models of musculoskeletal geometry (center) can be used to calculate the muscle–tendon lengths and velocities of the hamstrings during gait. The plots on the right compare the hamstring lengths and velocities of two subjects with crouch gait to the average of 45 unimpaired individuals (mean ±2 standard deviations). Adapted from Arnold et al. (2006).

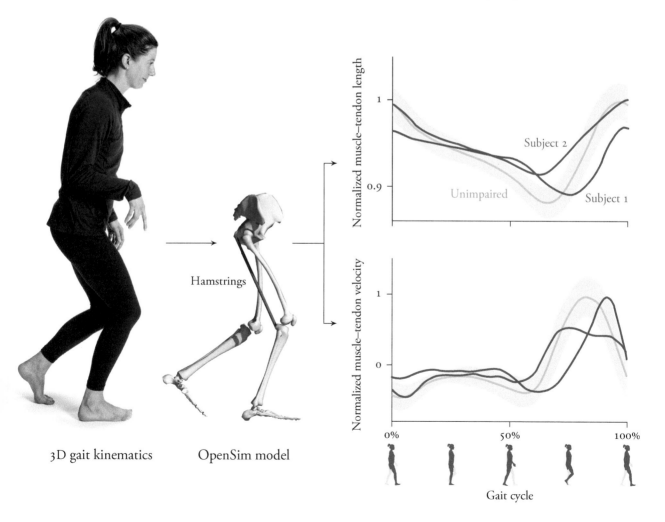

than normal. For example, Subject 1 walked with hamstring lengths that were substantially shorter than normal, while Subject 2 walked with hamstring velocities that were substantially slower than normal (Figure 6.10, right).

Analyzing hamstring lengths and velocities can indicate who might be a good candidate for hamstring surgery. Roughly one-third of the individuals we analyzed had hamstrings that were of normal length and normal velocity during walking, suggesting that their muscles were not too tight, and they would not benefit from surgery. About two-thirds of the individuals did have short or slow hamstrings; these patients were more likely to benefit from surgical lengthening of their hamstrings. While estimation of muscle lengths and velocities is only one piece of the puzzle, these data provide an objective basis for assessing which muscles should be surgically lengthened. To enable others to make these assessments, my laboratory provides free models of musculoskeletal geometry and open-source software, called OpenSim, for analyzing musculoskeletal mechanics. These tools are now used at many hospitals to help plan surgery, and you are welcome to use them as well.

Measurement and modeling of maximum joint moments

The previous sections illustrated that moment arms vary with joint angle, as do muscle forces. We also saw that a single muscle may actuate more than one joint. The reverse is also true—every joint is actuated by multiple muscles, each with its own architecture and geometry.

The muscles that actuate a joint come in pairs: *agonist* muscles that rotate the joint in one direction and *antagonist* muscles that rotate it in the opposite direction. For example, during walking, the tibialis anterior on the front of the leg generates force to control the lowering of the foot to the ground following foot contact (Figure 6.11, left). Muscles on the back of the leg subsequently generate force to plantarflex the ankle and generate a ground reaction force during the push-off phase of the gait cycle (Figure 6.11, right).

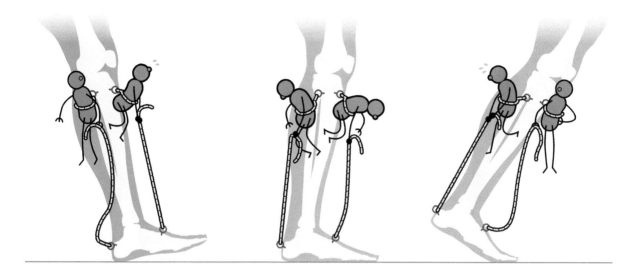

FIGURE 6.11
Skeletal muscles generate movement by pulling on the bones to which they attach. The forces in agonist and antagonist muscles are timed to produce coordinated motions. Inspired by Verne Inman's book *Human Walking* (1981).

The reason for this arrangement is simple: muscles only pull (they cannot push). This may seem counterintuitive at first, because in everyday speech we often talk about pushing (including in the last paragraph where we refer to the "push-off phase"). Yet when you carefully consider which muscles are doing the work when we are allegedly pushing, you will find that those muscles are indeed pulling. The "push-off" of the foot is actually generated by tension in the calf muscles.

Although only two muscles are illustrated in Figure 6.11, the agonist and antagonist muscles may in fact be groups of muscles. For instance, the so-called "quadriceps muscle" in the thigh actually consists of four separate muscles (the prefix "quadri-" means "four" in Latin). When predicting the maximum strength of the quadriceps, or of all the muscles acting on a joint in general, we need to take into account the architecture and geometry of each muscle, realizing that the total muscle moment will vary with joint angle and angular velocity.

The maximum voluntary isometric moment is often predicted from computer models of the musculoskeletal system, but it can also be measured experimentally using a dynamometer (Figure 6.12). For example, to measure the strength of the elbow flexor can be generated over the entire range of motion. In the clinic, one measures only the total joint moment (see Figure 6.12). Models and simulations are critical for understanding how individual muscles contribute to generating the measured joint moment.

Muscle architecture, moment arms, and tendon transfer surgery

The interplay between muscle architecture and moment arms in determining the functional capacity of a muscle can be illustrated by analyzing tendon transfer surgeries. These surgeries are performed on individuals with a nerve or spinal cord injury who have lost use of some of their muscles. In some cases, the tendon of a functioning muscle can be rerouted to the tendon of a nonfunctioning muscle to improve strength or range of motion. These surgeries may enable patients to use parts of their bodies that have been paralyzed. For example, a functioning elbow flexor muscle may be transferred to paralyzed wrist or finger muscles to restore the ability to grasp objects.

The role of moment arms in determining muscle moments is important in tendon transfer surgeries, in which the surgeon selects a muscle with an architecture that has evolved to function in one location and modifies its moment arms to achieve a new function. We must consider how the change in the muscle's moment arms will influence its ability to generate forces and moments about the joint of interest over the desired range of motion. Figure 6.14 provides a simplified illustration of a tendon transfer surgery. In this example, the surgeon has two choices for where to attach the functioning muscle, which has a maximum isometric force of 1.2 kN and an optimal fiber length of 8 cm. If she chooses to attach the muscle to the tendon of muscle A, the average moment arm of the newly configured muscle will be 4 cm (blue); if she chooses attachment site B, the average moment arm of the newly configured muscle will be 2 cm (orange).

Suppose the requirement is that the rerouted muscle must be able to generate a moment of at least 15 N·m over a 90-degree range of motion. Requirements like this one may be established so that the patient can achieve a specific goal, like bringing their hand to their mouth while holding a glass of water. Interestingly, whereas option A would generate a higher peak moment (48 N·m) due to the larger moment arm, the muscle fibers would also change length more and would therefore traverse a larger region of the active force–length curve (Figure 6.14, right). Because of the force–length relationship, option A would result in a minimum moment that is lower than the required 15 N·m at joint angles above about 76 degrees. By contrast, option B would result in a lower peak moment (24 N·m), but because the moment arm is smaller, the muscle would undergo a smaller change in length over the 90-degree range of motion. As such, the force and moment in option B would not vary much with joint angle, and the minimum moment would be around 21 N·m—well above the required 15 N·m. Thus, attachment site B would be the preferred choice in this scenario.

FIGURE 6.14
Application of muscle force and moment arm concepts to decision-making in tendon transfer surgery. In this example, transferring to the tendon of muscle A would generate a higher peak moment, but only option B would meet the given functional requirement.

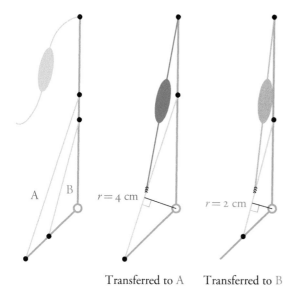

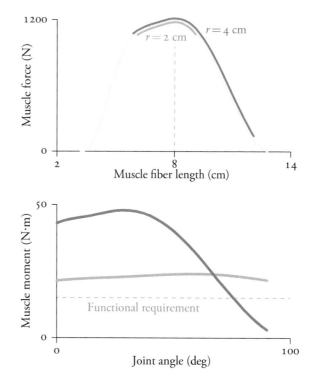

Moment arms of muscles with complex actions

The definitions provided above describe useful principles for understanding how muscle force leads to moment generation; however, there are several joints and muscles for which the assumptions described at the beginning of the chapter do not hold. In these scenarios, the muscle moment arms must be considered using a more sophisticated analysis. For example, if a muscle bulges appreciably as it contracts, then its moment arm depends on its active force. In this case, the assumptions we used to derive the tendon-excursion definition no longer hold, and therefore calculating $\partial \ell^{MT}/\partial \theta$ will no longer render the perpendicular distance from the muscle's line of action to the joint center (r; Equation 6.9). Similarly, if friction is present in the joint or as a muscle slides over other structures during joint rotation, then the assumption that muscles are frictionless and independent will be violated. Finally, there are several joints in which the position and orientation of the axis of rotation change during movement. In these situations, the moment arm must be calculated relative to the instantaneous center of rotation, which changes as the joint moves. Delp and Loan (1995) and Sherman et al. (2013) provide

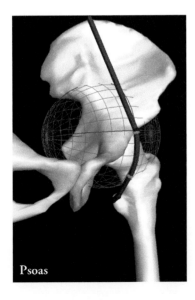
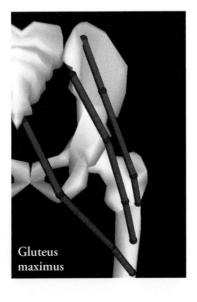

FIGURE 6.15
Representation of the psoas wrapping over the pelvic brim (left) and the gluteus maximus separated into three compartments (right). Adapted from Blemker and Delp (2005).

more in-depth descriptions of these special cases and how to handle them.

We must also represent the geometry of muscles that have broad attachments and muscles that wrap (curve) over bones and deeper muscles. The psoas muscle, for example, has a complex path that wraps around the pelvic brim, hip joint capsule, and femoral neck (Figure 6.15). Muscle paths such as this can be represented in a computer model using "via points" (the blue dots in the figure) or "wrapping surfaces" (wireframe), both of which prevent the muscle from penetrating adjacent structures. However, it is difficult to define via points and wrapping surfaces that work robustly for many degrees of freedom and over a large range of motion. The gluteus maximus is also difficult to model because it has broad areas of attachment and its fibers have a complex geometric arrangement. As shown in Figure 6.15, the muscle can be divided into three muscle–tendon compartments to better approximate its mechanical function. An alternative to via points and wrapping surfaces is to represent the three-dimensional fiber geometries with dozens or hundreds of paths (Figure 6.16). The appropriate level of modeling complexity depends on the goal of the analysis. For

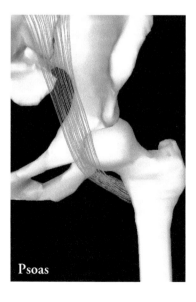
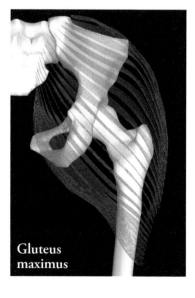

FIGURE 6.16
Representations of fiber geometries of the psoas (left) and gluteus maximus (right) allow for detailed modeling of muscle geometry and force generation. Adapted from Blemker and Delp (2005).

example, to estimate the total moment generated by muscles with relatively simple geometry, a model like that shown in Figure 6.6 is likely to be sufficient. In contrast, to calculate how the variation of fiber lengths within a muscle affects its force-generating capacity, more detailed models like those shown in Figure 6.16 may be required.

Wrapping up

We have seen how a muscle's moment arm affects its ability to generate forces and joint moments. Thus far, we have assumed that the muscles are maximally activated, as they would be during a strength assessment; however, muscles are rarely maximally activated during movement. To estimate the forces and moments generated by muscles during submaximal activation, we might use the process illustrated in Figure 6.17. In the example shown, electromyographic (EMG) data are collected experimentally, filtered, and used to drive a model of activation dynamics (Chapter 4). Each muscle activation is then used as an input to a model of muscle contraction dynamics (Chapter 5). The other key inputs to the contraction dynamics model—namely, the length and velocity of the muscle–tendon unit—can be determined from measurements of joint angles and a model of the musculoskeletal system.

The process shown in Figure 6.17 describes a type of forward problem: given EMG data and joint angles, find the resulting muscle forces and joint moments. We will use exactly this formulation in Chapter 12 to study the energy stored in tendon during running. Biomechanists also frequently solve inverse problems in which effects are observed and then algorithms are used to determine the underlying phenomena responsible for inducing these effects. For example, we often measure motions of the body and wish to estimate the muscle forces that must have been present to produce the observed movement.

In the next three chapters, we will leverage what we have learned thus far and explore some of the tools that are used for inverse analyses. In Chapter 7, we will learn how computational

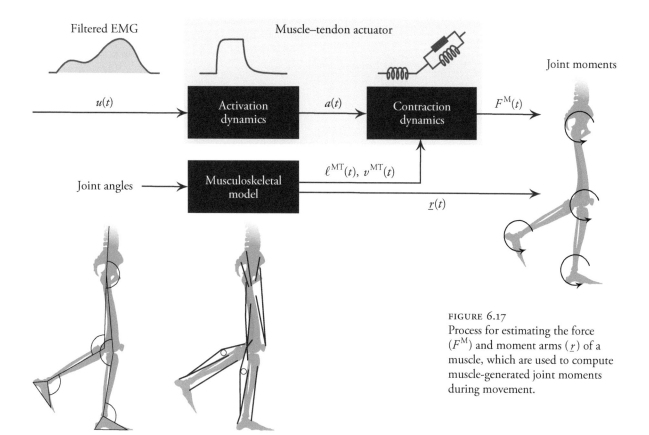

FIGURE 6.17
Process for estimating the force (F^M) and moment arms ($\underline{r}$) of a muscle, which are used to compute muscle-generated joint moments during movement.

models and algorithms are used to estimate joint angles from measurements of body segment motions. Chapter 8 describes how we can use these joint angles along with measurements of external forces to estimate net joint moments. Finally, in Chapter 9, we will use optimization to estimate the muscle forces that would have generated the net joint moments and elicited the joint trajectories that we calculated. By viewing just a few noninvasive measurements through the lens of a musculoskeletal model, we can calculate internal joint forces, muscle activity, energy expenditure, and other valuable information.

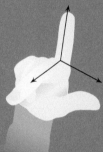

Part III
Analysis of Movement

7 Quantifying Movement

All change is not growth,
as all movement is not forward.
—Ellen Glasgow

IT'S THE MOMENT that every soccer player dreads. An awkward landing, a popping sound. Excruciating pain in the knee and crumpling to the ground as the knee can no longer support any weight. An examination by the team doctor, and the three letters no one wants to hear in this situation: ACL. As in, anterior cruciate ligament, the most common site of knee injuries.

Perhaps worse than the initial pain is what comes next: an operation in which a surgeon drills holes through the tibia and femur, weeks of wearing a brace, followed by months of rehabilitation. After all this, the chances of returning to action are good. But returning to action is not the same as living happily ever after. Players who return from an ACL injury have a greater risk of reinjury and an elevated risk of developing arthritis in the injured knee. Long after the competition is over, the athlete will still be paying the price.

What if ACL injuries could be prevented? The majority of ACL injuries are not caused by contact, and many of them might be prevented with training exercises that focus on balance, proper landing technique, and improved awareness of body position. The training can be even more effective when targeted at athletes who are most at risk. Measuring an athlete's motion can determine whether they are at risk of injuring their ACL. Do they land from a jump with their knees aligned (good) or misaligned (bad)? Do they absorb the energy of impact by landing with flexed knees (good) or are their knees more extended (bad)?

At high levels of athletic competition, details matter, and there is no substitute for quantitative measures of performance. The same is true in scientific research. Lord Kelvin, the famous mathematical physicist after whom the unit of absolute temperature was named, had this to say about quantification (Thomson, 1889, pp. 73–74):

> I often say that when you can measure what you are speaking about, and express it in numbers, you know something about it; but when you cannot measure it, when you cannot express it in numbers, your knowledge is of a meagre and unsatisfactory kind: it may be the beginning of knowledge, but you have scarcely, in your thoughts, advanced to the stage of science, whatever the matter may be.

In this chapter, we will introduce the most common techniques for measuring movement, including force plates and motion capture using cameras. We also describe techniques for determining the pose of a subject's skeleton, known as the *inverse kinematics* problem.

Our ultimate goal is to understand the muscle forces that produce movement, but these forces usually cannot be measured directly. Instead, we estimate these muscle forces by constructing a skeletal model and computing the joint angles over time that best agree with the observed motion (this chapter). We then compute the forces and moments acting at each joint (Chapter 8), and finally solve an optimization problem that reflects how the body coordinates dozens of muscles to generate these forces and moments (Chapter 9). Clearly, we must have a solid foundation to achieve the more ambitious goals of the next two chapters. The quality of all our subsequent analyses depends on how accurately we compute the joint angles, which is our main objective in this chapter. But first, let's take a step back in time to see how inverse kinematics methods developed.

Measurement techniques

Early studies of movement were limited to qualitative observations and measurements of gross body motion. Advances in the photographic process in the 1800s made photography practical and enabled exciting new investigations of movement. In 1872, Leland Stanford (who, with his wife Jane, founded Stanford

University) sought to settle an argument about the locomotion of horses. When trotting, a horse advances diagonally opposite legs in unison. Stanford believed there was a period of time during which all four hooves of a trotting horse were airborne simultaneously and hired Eadweard Muybridge to investigate. Muybridge and engineer John Isaacs pioneered a technique for photographing a horse during locomotion using a series of cameras positioned along the horse's path, and were able to capture the instant a trotting horse was airborne. The *Horse in Motion* photographs (Figure 7.1) revolutionized the field of biomechanics and launched the development of motion pictures.

FIGURE 7.1
Yea or *neigh*? In 1872, Leland Stanford hired Eadweard Muybridge to determine whether all four hooves are airborne simultaneously when a horse trots. The fourth photograph in this sequence captures an instant at which all four hooves are airborne. Image courtesy of Stanford University.

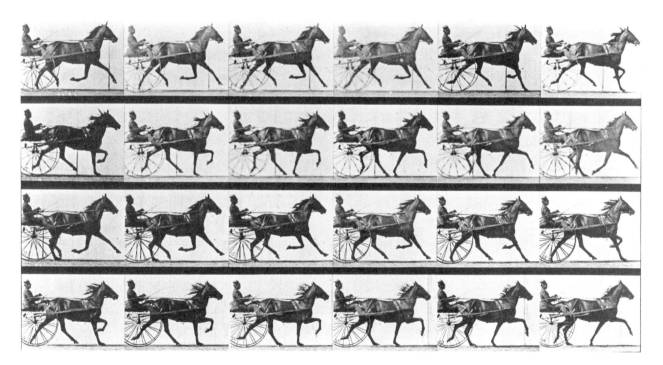

While video analysis remains a key component of biomechanics, other techniques have been developed for measuring movement. For example, submillimeter measurements of bones can be obtained using fluoroscopy, in which X-ray images are taken continuously over time (Figure 7.2). Unfortunately, the exposure to ionizing

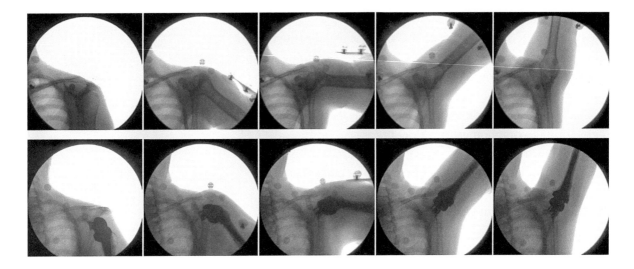

FIGURE 7.2
Fluoroscopic images showing bone motions in a healthy shoulder (top row) and following a total shoulder replacement (bottom row). Images courtesy of the University of Utah Orthopaedic Research Laboratory.

FIGURE 7.3
Inertial measurement units, the orange sensors on the runner's pelvis and ankles, can measure linear accelerations and angular velocities throughout long runs to assess musculoskeletal loading.

radiation (as well as a limited field of view) often makes this technique impractical for large-scale studies. Accurate kinematic measurements can also be obtained by implanting bone-anchored pins that protrude from the skin and tracking the motion of these pins using optical techniques. As you might expect, the invasiveness of this procedure makes it impractical for most research studies.

Inertial measurement units (IMUs) have become popular as a low-cost alternative to existing strategies for quantifying movement. IMUs typically comprise accelerometers for measuring linear accelerations, gyroscopes for measuring angular velocities, and a magnetometer for measuring heading relative to magnetic north. In many studies, IMUs eliminate the need for cameras and controlled lighting conditions, which enables one to perform biomechanical experiments in natural environments, during activities like swimming and trail running, and over long durations to capture infrequent events or to monitor health (Figure 7.3). For example, Helen Bronte-Stewart used IMUs to measure the angular

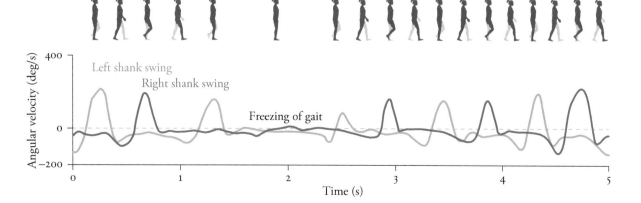

FIGURE 7.4
Inertial measurement units can measure the angular velocity of the shank in the sagittal plane (top), detecting freezing-of-gait episodes as individuals with Parkinson's disease navigate living environments (bottom). Adapted from O'Day et al. (2020).

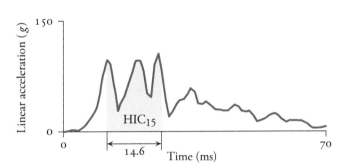

velocity of each leg in the sagittal plane as individuals walked through a simulated living environment, where optical motion capture would be impractical. The beginning and end of each swing phase corresponded to zero-crossings of the angular velocity signals they measured. This experimental setup enabled them to detect freezing-of-gait episodes in individuals with Parkinson's disease (Figure 7.4), a gait disturbance in which walking is suddenly interrupted. IMUs have also been used to measure the motion of a football player's skull (Figure 7.5) and could be used to help monitor the health of an athlete's brain.

Optical motion capture

Motion capture, or "mocap," has evolved over nearly a century. Besides its role in biomechanics, it has also been an integral part of the entertainment industry. To animate characters in movies and video games, we go from reality (an actor swinging a stick around the room) to a series of images (an alien catching a butterfly in a net). In biomechanics, we are, in a sense, reversing this process: we use a series of images to determine what the underlying bones and muscles were doing to produce an observed motion.

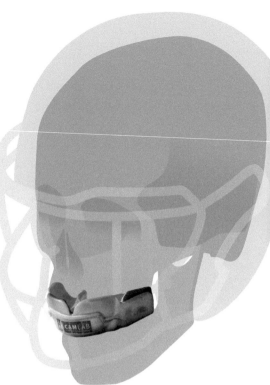

FIGURE 7.5
Kinematic measurements for monitoring athlete health. A mouthguard instrumented with an inertial measurement unit provides measurements of skull motion. The Head Injury Criterion (HIC) is proportional to the area under the total acceleration curve. In this example, the safe threshold for a 15-millisecond time window is accumulated in only 14.6 milliseconds, leading to a traumatic brain injury. Adapted from Hernandez et al. (2015).

Whether in a movie studio or a biomechanics lab, the technology is the same. We affix small spherical retroreflective markers either to the subject's skin or onto plates that are then attached to the subject's body (Figure 7.6). Video cameras track the motion of these markers, typically using infrared light. Operating in the infrared spectrum has two benefits: it reduces the effects of ambient light and allows for pulsed illumination to reduce motion-blur artifacts. The stroboscopic effect would be very distracting if we used visible light instead.

In general, three noncollinear markers must be placed on a body segment to determine its position and orientation in space. Fewer than $3n$ markers can track n body segments if some markers are assumed to be common to adjacent segments (e.g., if the marker on the lateral side of the knee is assumed to remain fixed relative to both the thigh and the shank). It is also possible to reduce the number of markers by using a kinematic model of the underlying skeleton. However, using more markers may improve joint angle estimates for some inverse kinematics algorithms.

Each camera records a series of 2D images, capturing the location of each marker in that camera's local 2D image plane over time (Figure 7.7). For each time frame, a computer combines the location information from all cameras with knowledge of the relative position and orientation of each camera in space (determined during an earlier calibration procedure) to estimate the 3D location of each marker relative to a "global" laboratory reference frame. Because a single image does not provide accurate information about the distance between the camera and a marker, each marker must be visible by at least two cameras to determine its location in space. Increasing the number of cameras in the lab and the distance between them can improve marker location estimates, account for temporary marker occlusion that occurs as the subject moves, and enlarge the volume through which the subject's movement can be measured. Two to four cameras may be sufficient for two-dimensional analyses; three-dimensional analyses over a large volume may require dozens. When a mocap system is calibrated and operated correctly, marker measurement errors are no greater than a few millimeters in a typical biomechanics experiment.

FIGURE 7.6
Optical motion capture (mocap) is a popular technique for quantifying movement. We can estimate the motion of the underlying skeleton from the trajectories of points on the skin.

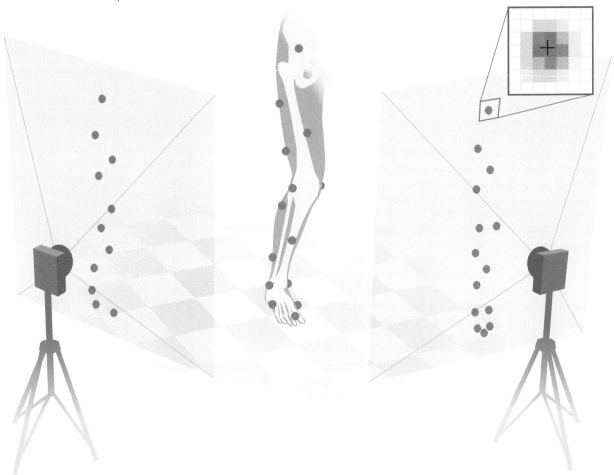

The largest source of experimental measurement error is the motion of soft tissue. Muscles bulge, fat jiggles, and skin stretches as we move, changing the relationship between the skin-mounted markers and the underlying bones we wish to track. Soft-tissue artifacts are unavoidable, but their effects can be reduced by placing markers in strategic locations and by using algorithms that account for these artifacts. Several marker placement protocols are in common use; selection of a suitable protocol depends, in part, on the objectives of the study being performed and the inverse kinematics algorithm being used. Markers are often placed on bony anatomical landmarks where soft-tissue motion is relatively low, such as the tibial and fibular malleoli at the ankle.

FIGURE 7.7
The location of each marker is determined in each camera's local 2D image plane (e.g., by computing the centroid of each region of illuminated pixels, weighted by intensity; see crosshairs in inset). These 2D locations are then combined with information obtained during camera calibration to calculate the best estimate for the location of each marker in space, relative to a global reference frame.

Placing markers on anatomical landmarks may be impractical in locations with large amounts of fat or because the marker would interfere with the motion being studied (e.g., on the medial side of the knee during walking). In such cases, "virtual markers" can be defined by relating the positions of anatomical landmarks (for example, the greater trochanter) to the positions of nearby skin-mounted markers. When studying obese subjects, for example, we press a calibration wand of known length against the skin until the tip of the wand reaches the anatomical landmark. We capture a single frame of video to calculate the location of the tip from the positions of markers mounted along the length of the wand. This information gives us a relationship between the position of the markers on the skin and the location of the anatomical landmark, which we can then use to estimate the trajectory of the anatomical landmark during motion trials. It can also be useful to define virtual markers at joint centers, the locations of which can be estimated using the trajectories of skin-mounted markers as the joint moves through a large range of motion.

Even after we have created virtual markers, we typically need to do more work before our video data are suitable for analysis. For example, we often want to obtain velocity data by taking the derivative of the measured position data. But differentiation amplifies the high-frequency noise that these data typically contain. One way to avoid this is to smooth the data with a low-pass filter, which removes high-frequency motion. But we must be careful to avoid filtering out meaningful information. Because the frequency range of interest depends on the activity being studied, so too does the appropriate filter cutoff frequency. For example, we may be interested in frequencies up to about 3 Hz for studying posture, 6 Hz for studying walking, and 10–15 Hz for studying running. By contrast, frequencies up to 300 Hz may be relevant in a study of rapid changes in motion during the heel strike of running. Unfortunately, soft-tissue artifacts often appear in the same frequency range as the signal of interest, so we need to address these errors through strategic marker positioning and algorithms that are designed to take them into account.

There are two other ways in which we process the raw video data before analyzing it. First, it is common for markers to become temporarily obscured from a camera's view during an experiment, and thus the measured trajectories of these markers are incomplete. Short missing segments may be approximated using interpolation; longer gaps may be accommodated by simply ignoring the occluded marker, provided that enough other markers are visible. Second, the cameras cannot distinguish between the retroreflective optical markers; a label must be applied to each marker in each frame of video. We may need to assign these labels manually or correct errors that were made by an automated labeling method. To avoid marker labeling issues altogether, light-emitting "active" markers can be used in place of the retroreflective "passive" markers described above. Active markers can be sequentially pulsed so that only one marker is illuminated at any one time, making them uniquely identifiable. Despite this advantage, we seldom use active markers in my laboratory. Many systems require a power supply

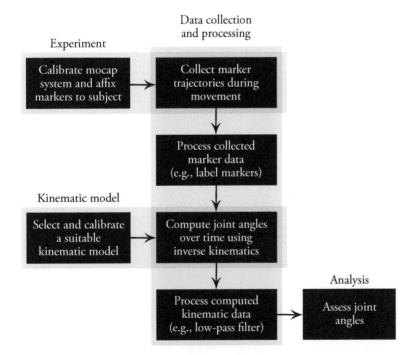

FIGURE 7.8
Typical process for computing joint angles from mocap data. The trajectories of skin-mounted markers are recorded by a calibrated motion capture system. A kinematic model of the subject is then used to estimate skeletal joint angles from 3D marker positions.

and cables, which can encumber a subject's motion, and the frequency at which each marker can be pulsed (the rate at which data can be collected) decreases as more markers are added. For example, a mocap system operating at 1 kHz can capture only 10 markers unambiguously if they are pulsed at 100 Hz. A typical process for computing joint angles from mocap data is summarized in Figure 7.8 and described below.

From Muybridge to the modern day, video-based analysis has been the workhorse of movement research. In the future, markerless motion capture technologies that use only video will increase in accuracy and popularity. Body-worn accelerometers and IMUs will dramatically expand the types of biomechanical studies that can be performed and increase the number of people who participate in them. In the remainder of this chapter, however, we will focus on the most common techniques currently used in biomechanics research and clinical motion capture labs, where we collect optical marker trajectories and compute skeletal joint kinematics.

Unconstrained inverse kinematics

One common method for estimating joint angles from marker trajectories involves first establishing a reference frame (Figure 7.9) fixed to each body segment based on the positions of the markers attached to that segment. The relative orientations of reference frames fixed to adjacent segments are then interpreted as the angles of the joints connecting the segments (Figure 7.10). We refer to this method as "unconstrained" inverse kinematics because no constraints, or limits, are imposed on either the limb lengths or the joint motions of the implied skeletal model. (This approach has also been referred to as the "direct method" because joint angles are calculated "directly" from marker positions. We discourage using this term to prevent confusion with "direct kinematics," a synonym for "forward kinematics," which describes the opposite process of computing the positions of markers given the angles of the joints.) Note that we are often interested in the joints' angular velocities and angular accelerations as well, but we will assume that the joint

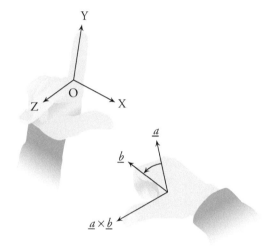

FIGURE 7.9
A reference frame fixed to a body is defined by a point on that body (the "origin" of the reference frame, O) and three orthogonal axes (X, Y, and Z). By convention, we use reference frames that are "right-handed": if you extend the thumb and index finger of your right hand and align them with the X and Y axes, your middle finger will be aligned with the Z axis when it is pointed away from your palm (left). If you curl the fingers of your right hand from $\underline{a}$ to $\underline{b}$, your fully extended thumb points in the direction of $\underline{a} \times \underline{b}$ (right). In a right-handed coordinate system, $\hat{x} \times \hat{y} = \hat{z}$.

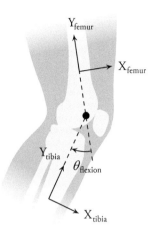 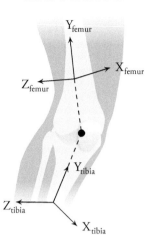 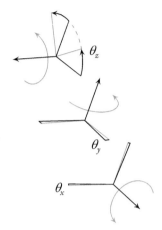

angles can be differentiated with respect to time (after filtering, if necessary) to obtain these quantities.

Suppose we sought to measure a subject's knee flexion angle (specifically, the angle between the femur and tibia about the knee joint axis) throughout a gait cycle. In the simplest scenario, we would track the motion of two reference frames over time: one fixed to the femur and one fixed to the tibia, both aligned with the anatomical knee joint axis. The knee angle could then be defined simply as the relative orientation of these reference frames in the plane orthogonal to the knee axis (Figure 7.10). More commonly, however, we would track the positions of markers affixed to the thigh and shank, and must perform two additional calculations. First, we must compute the position and orientation of the thigh and shank body segments by constructing body-fixed reference frames from the marker positions. Second, because markers placed on the medial femoral epicondyles are likely to be occluded and may interfere with the subject's gait, the location and orientation of the knee anatomical joint axis is generally not tracked directly and must also be determined from the positions of the markers affixed to the thigh and shank. Clearly, this scenario is more complex, and the remainder of this section describes each step of the unconstrained inverse kinematics method in detail.

FIGURE 7.10
Joint angles can be calculated by comparing the orientations of reference frames fixed to adjacent body segments. Knee flexion can be computed by inspection in a two-dimensional analysis (left), but a more formal approach is useful when computing the relative orientation between two frames in three dimensions (right).

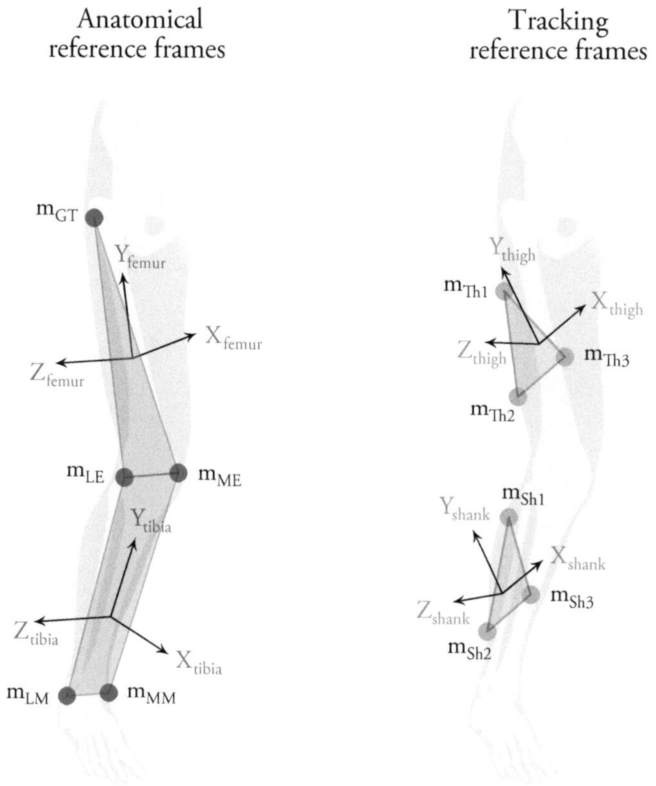

QUANTIFYING MOVEMENT 173

Anatomical reference frames

Tracking reference frames

FIGURE 7.11
Body-fixed reference frames determined from markers mounted on anatomical landmarks (anatomical reference frames) and markers mounted on body segments (tracking reference frames); m = marker, GT = greater trochanter, ME/LE = medial/lateral femoral epicondyle, MM/LM = medial/lateral malleolus, Th = thigh, Sh = shank.

The number and locations of markers can vary depending on the study objectives, the motion being tracked, and the software used for analysis. It is common to use two sets of markers, one to define anatomical reference frames and one to define tracking reference frames (Figure 7.11). Anatomical reference frames represent the underlying skeletal structure and are defined by placing markers on anatomical landmarks. For example, the anatomical reference frame fixed to the femur might be defined as follows:

1. The origin is midway between the greater trochanter marker (m_{GT}) and the knee joint center (the midpoint between the medial and lateral femoral epicondyle markers, m_{ME} and m_{LE});
2. Z_{femur} is parallel to the knee joint axis, which is defined as the vector from medial to lateral femoral epicondyle markers, normalized to unit length;

3. X_{femur} is the cross product of Z_{femur} and a vector from the greater trochanter marker to one of the femoral epicondyle markers, normalized to unit length;
4. Y_{femur} is the cross product of Z_{femur} and X_{femur}, which completes the right-handed reference frame (Figure 7.9).

A reference frame fixed to the tibia can be defined similarly, again using the femoral epicondyle markers to define the Z axis. We can then define the knee flexion angle as the angle of the tibia reference frame relative to the femur frame about the (parallel) Z axes.

Tracking reference frames are also fixed to body segments, but they might not be aligned with anatomically relevant axes. Tracking reference frames are defined by at least three noncollinear markers on each segment. These markers need not be distinct from those used to define anatomical reference frames, but they should be placed in locations that will remain visible throughout the experiment and have relatively small amounts of soft-tissue motion. For example, the greater trochanter and lateral femoral epicondyle markers could be used to define the thigh reference frame along with one additional marker placed somewhere on the thigh. An orthogonal, right-handed reference frame can then be defined from these markers. Tracking reference frames are useful because they allow us to place markers in more convenient locations than the anatomical landmarks. We still require anatomical reference frames to compute joint angles, but the anatomical frames can be readily computed once the tracking frames are known, as described in the next section.

Transformation matrices

In the example above, computing the knee flexion angle was straightforward because the anatomical reference frames fixed to the femur and tibia were defined with parallel Z axes. Thus, the orientation of one frame relative to the other could be described by a single angle of rotation about the common Z axis. In general, we relate two arbitrarily oriented reference frames by not one but *three* angles, which we call Euler angles (named after the eighteenth-century mathematician Leonhard Euler). These angles always describe a sequence of consecutive rotations about three axes, but the order of the rotations is not unique. For example, you might describe rotating about the X axis, then about the Y axis, and finally about the Z axis. Alternatively, you might rotate about the Z axis first, or about the rotated Y axis rather than the original one. A few choices for these three axes are more common than the others. In aeronautics, for example, the axes of rotation are traditionally chosen such that the three angles are the yaw, pitch, and roll (Figure 7.12). This sequence of rotations is called a Z–Y–X body-fixed rotation sequence, "body-fixed" because each rotation occurs with respect to the frame that is attached to the body. (The alternative is called "space-fixed," where each rotation is performed about one of the lab frame's axes.)

Suppose we are given two reference frames, A and B, and a point whose coordinates are expressed relative to frame B, and we wish to compute this point's coordinates relative to frame A. We can relate the orientations of these two frames by means of a *rotation matrix*, a 3×3 matrix $^{A}R_{B}$ whose columns are simply the coordinates of unit vectors pointed along frame B's X, Y, and Z axes, when they are expressed relative to frame A. Once we know this matrix, we can transform the frame-B coordinates of any point, p_{B}, into the corresponding frame-A coordinates by multiplying the matrix $^{A}R_{B}$ by the vector p_{B}:

$$p_{A} = {}^{A}R_{B}\, p_{B} \qquad (7.1)$$

By analogy, we can transform the frame-A coordinates of any point, p_{A}, into the corresponding frame-B coordinates by multiplying the matrix $^{B}R_{A}$ by the vector p_{A}. Equivalently, we could compute p_{B} from p_{A} by premultiplying each side of Equation 7.1 by $(^{A}R_{B})^{-1}$:

$$\left({}^{A}R_{B}\right)^{-1} p_{A} = p_{B} \qquad (7.2)$$

which reveals the following relationship:

$$^{B}R_{A} = \left({}^{A}R_{B}\right)^{-1} \qquad (7.3)$$

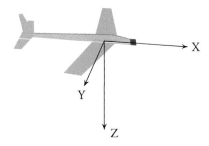

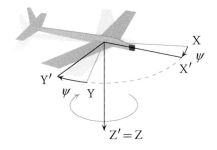

Yaw:
Rotate by ψ
about Z

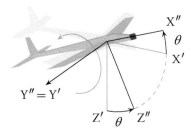

Pitch:
Rotate by θ
about Y'

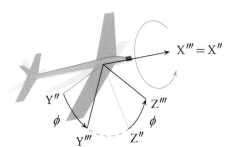

Roll:
Rotate by ϕ
about X''

FIGURE 7.12
Yaw, pitch, and roll describe the three angles of rotation in the popular Z–Y–X body-fixed rotation sequence.

In fact, rotation matrices satisfy a property called orthogonality, which means that the inverse of a rotation matrix is equal to its transpose:

$$\left(^A R_B\right)^{-1} = \left(^A R_B\right)^T \tag{7.4}$$

Thus, we can also define $^A R_B$ as the rotation matrix whose rows are the coordinates of unit vectors pointed along frame A's X, Y, and Z axes, when they are expressed relative to frame B.

To derive the relationship between the matrix $^A R_B$ and a particular set of Euler angles, we will begin with an analogy. Imagine that you are sitting in an airplane preparing for takeoff and have defined a coordinate system whose X axis points out the nose of the plane, whose Y axis points toward the right wing, and whose Z axis points down (Figure 7.12). We will call this frame the lab frame. Now, fasten your seatbelt and imagine the plane executing three maneuvers. First, it turns to the right as it taxis toward the runway, rotating about the Z axis by an angle ψ. (For now, we will ignore the distance the plane has moved down the taxiway and consider only the change in its orientation.) This maneuver, called a "yaw" in aeronautics, results in a new set of basis vectors for axes X′, Y′, and Z′. Note that Z′ = Z because the direction of the "down" vector has not changed as a result of the right turn. The axes {X′, Y′, Z′} define a new reference frame, which we will call frame 1 or f1.

Second, the plane takes off, lifting its nose (rotating about the Y′ axis) by an angle θ. This motion, which the pilot would call "pitch," results in another new reference frame, f2 = {X″, Y″, Z″}, as shown in Figure 7.12. This time, Y″ = Y′ because pitching the nose upward does not affect the direction in which the right wing is pointing.

Third, the plane banks to one side, tipping the right wing down (rotating about the X″ axis) by an angle ϕ. This maneuver is called "roll" in aeronautics and results in a third reference frame that we will call the body frame: body = {X‴, Y‴, Z‴}. In this case, X‴ = X″ because tipping a wing up or down does not change the direction in which the plane's nose is pointing. In this manner, we can arrive at the reference frame fixed to any body in space, starting from an inertial or lab-fixed frame and performing the above three maneuvers in sequence.

How do we transform from a (movable) body-fixed reference frame back to a (stationary) lab-fixed frame? In symbols, what is the matrix $^{lab}R_{body}$? Or in words, how do we convert a measurement from a body-fixed sensor to an observation in our preferred lab coordinates?

Thanks to our airplane analogy, we hope that the answer is relatively obvious: we undo the above sequence of rotations. To transform from body to lab coordinates, we first transform from the body frame to the f2 frame. Then we transform from the f2 frame to the f1 frame. Finally, we transform from the f1 frame to the lab frame. Thus, we compute the rotation matrix $^{lab}R_{body}$ as follows:

$$^{lab}R_{body} = {}^{lab}R_{f1} \, {}^{f1}R_{f2} \, {}^{f2}R_{body} \qquad (7.5)$$

Note the order of the transformations, which may not be obvious at first. Because these transformations can be thought of as a sequence of functions, we write their composition from right to left, so transformation $^{f2}R_{body}$ (from body coordinates to f2 coordinates) is applied first.

A big payoff of the above decomposition is that each of the three rotation matrices has a particularly simple form. We start with the transformation from f1 coordinates to lab coordinates:

$$^{lab}R_{f1} = \begin{bmatrix} \cos\psi & -\sin\psi & 0 \\ \sin\psi & \cos\psi & 0 \\ 0 & 0 & 1 \end{bmatrix} \qquad (7.6)$$

To see why the matrix has this form, remember that the first column of a rotation matrix expresses the coordinates of the new X' axis with respect to the original {X, Y, Z} reference frame. And remember what X' is: the direction the nose of the plane points after rotating about the Z axis by the angle ψ. A unit vector in that direction is $(\cos\psi, \sin\psi, 0)$, which are indeed the entries in the first column of the matrix. Similarly, the third column is the direction of the Z' axis in the {X, Y, Z} reference frame. Keeping in mind that Z' = Z, as we pointed out earlier, we conclude that the unit vector in this direction is just (0, 0, 1), and hence that is the third column of the rotation matrix.

We can also check this matrix by transforming points along each coordinate axis in frame f1 to the lab frame. For example, a bird sitting on the right wing might have the following coordinates when expressed with respect to the lab frame:

$$p_{lab} = {}^{lab}R_{f1}\, p_{f1} = \begin{bmatrix} \cos\psi & -\sin\psi & 0 \\ \sin\psi & \cos\psi & 0 \\ 0 & 0 & 1 \end{bmatrix} \begin{Bmatrix} 0 \\ 1 \\ 0 \end{Bmatrix} = \begin{Bmatrix} -\sin\psi \\ \cos\psi \\ 0 \end{Bmatrix} \quad (7.7)$$

If the airplane were to make a right turn ($\psi = 90$ degrees), then the bird would end up where the rudder used to be. The bird's coordinates relative to the original frame would be (−1, 0, 0). Using similar arguments, we can show that

$$^{f1}R_{f2} = \begin{bmatrix} \cos\theta & 0 & \sin\theta \\ 0 & 1 & 0 \\ -\sin\theta & 0 & \cos\theta \end{bmatrix} \quad (7.8)$$

and

$$^{f2}R_{body} = \begin{bmatrix} 1 & 0 & 0 \\ 0 & \cos\phi & -\sin\phi \\ 0 & \sin\phi & \cos\phi \end{bmatrix} \quad (7.9)$$

The matrices in Equations 7.6, 7.8, and 7.9 are called *elementary rotation matrices* because, in each case, one of the body-fixed coordinate axes remains stationary. The full rotation transformation $^{lab}R_{body}$ is given by the product of these three elementary rotation matrices in the prescribed order:

$$\begin{aligned}
^{lab}R_{body} &= {}^{lab}R_{f1}\, {}^{f1}R_{f2}\, {}^{f2}R_{body} \\
&= \begin{bmatrix} \cos\theta\cos\psi & \sin\phi\sin\theta\cos\psi - \cos\phi\sin\psi & \cos\phi\sin\theta\cos\psi + \sin\phi\sin\psi \\ \cos\theta\sin\psi & \sin\phi\sin\theta\sin\psi + \cos\phi\cos\psi & \cos\phi\sin\theta\sin\psi - \sin\phi\cos\psi \\ -\sin\theta & \sin\phi\cos\theta & \cos\phi\cos\theta \end{bmatrix} \\
&= \begin{bmatrix} r_{11} & r_{12} & r_{13} \\ r_{21} & r_{22} & r_{23} \\ r_{31} & r_{32} & r_{33} \end{bmatrix} \quad (7.10)
\end{aligned}$$

where each entry r_{ij} is a number between −1 and 1. We can compute the angles ψ, θ, and ϕ from the numeric entries of any rotation matrix by setting its entries equal to the corresponding trigonometric expressions in Equation 7.10. A unique solution is obtained only if limits are imposed on the three angles; one strategy is to use the 4-quadrant (2-argument) inverse tangent function (atan2) as follows:

$$\phi = \operatorname{atan2}(r_{32}, r_{33})$$
$$\theta = \operatorname{atan2}\left(-r_{31}, \sqrt{r_{11}^2 + r_{21}^2}\right) \quad (7.11)$$
$$\psi = \operatorname{atan2}(r_{21}, r_{11})$$

Note that we could compute the angles corresponding to a different sequence of rotations (e.g., rotating about X, then Y′, then Z″) simply by relating the numeric entries r_{ij} to a different symbolic rotation matrix. We would simply compute the product of a different set of elementary rotation matrices in Equation 7.10.

In the foregoing discussion, we assumed that all reference frames shared a common origin and thus were related through rotation transformations alone. However, reference frames that are attached to different body segments (e.g., the thigh and shank) usually have different origins and thus are related by translations as well as rotations. In general, the transformation from coordinates in frame B to coordinates in frame A is accomplished by a combination of a rotation $^A R_B$ and a translation:

$$p_A = {}^A R_B\, p_B + p_A^{AoBo} \quad (7.12)$$

where p_A^{AoBo} is the position vector from the origin of frame A (Ao) to the origin of frame B (Bo), expressed in frame A. (Note that Equation 7.12 simplifies to Equation 7.1 when the origins of frames A and B are coincident.) This kind of transformation, which combines rotations with a translation, can be conveniently represented using a 4×4 transformation matrix that captures the relative position and orientation between two frames:

$$^A T_B = \begin{bmatrix} & & & | & \\ & ^A R_B & & | & p_A^{A_o B_o} \\ & & & | & \\ \hline 0 & 0 & 0 & | & 1 \end{bmatrix} \quad (7.13)$$

Thus, a point expressed in frame B can be expressed in frame A using a transformation matrix as follows:

$$\begin{Bmatrix} p_A \\ 1 \end{Bmatrix} = {}^A T_B \begin{Bmatrix} p_B \\ 1 \end{Bmatrix} \quad (7.14)$$

Equation 7.14 effectively includes the offset vector $p_A^{A_o B_o}$ on the right-hand side of Equation 7.12 while preserving the notational convenience of relating two frames with a single matrix ($^A T_B$).

Finally, as a note to readers who are particularly interested in the mathematics, we will mention that representing rotations with rotation matrices (e.g., Equation 7.10) is not always the best choice. This method suffers from a defect called "gimbal lock": it becomes impossible to represent certain rotations when two or more axes of rotation are aligned. For example, if pitch $\theta = -90$ degrees in Figure 7.12, then $Z = X''$, and changing angles ψ and ϕ would have the same effect on the airplane. Gimbal lock can be avoided in applications where the coordinate frames have a limited range of motion. But if, for example, you are programming a flight simulator and you want users to be able to execute smooth somersaults, then transformation matrices may be a poor choice. Another representation, called the quaternion, avoids gimbal lock but at the cost of using more esoteric mathematics.

Calculating joint angles with unconstrained inverse kinematics

The great advantage of transformation matrices is that they simplify navigation among multiple coordinate systems. In particular, they allow us to easily calculate anatomical joint angles from data collected in the laboratory frame. Recall that a mocap system records the positions of skin-mounted markers expressed relative to a reference frame fixed in the lab (Figure 7.13). To calculate the knee

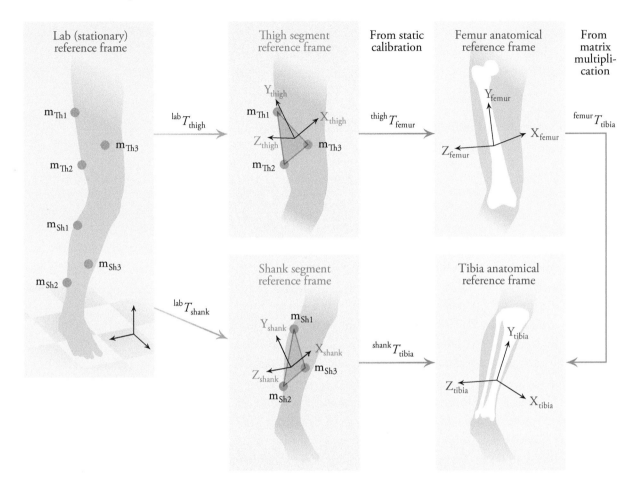

flexion angle, for example, we can determine the transformation matrix relating the reference frames fixed to the femur and tibia:

$$^{femur}T_{tibia} = {}^{femur}T_{thigh}\,{}^{thigh}T_{lab}\,{}^{lab}T_{shank}\,{}^{shank}T_{tibia} \quad (7.15)$$

Reading this formula from right to left describes how coordinates in the tibia frame are related to coordinates in the femur frame.

Assuming there is neither bone deformation nor soft-tissue motion, the thigh and shank tracking reference frames can be related to the femur and tibia anatomical frames, respectively, through constant transformations. These transformation matrices ($^{thigh}T_{femur}$ and $^{shank}T_{tibia}$) can be obtained by relating the tracking and anatomical reference frames during a "static calibration trial"

FIGURE 7.13
The relative position and orientation of any two reference frames can be described by a 4×4 transformation matrix T, which facilitates calculation of joint angles from marker locations measured relative to a global reference frame.

in which both sets of markers are attached to the subject and data are collected during standing. Transformations $^{lab}T_{thigh}$ and $^{lab}T_{shank}$ are obtained by computing thigh and shank reference frames from the marker positions recorded by the mocap system at each instant of the motion trial. The knee flexion angle can then be extracted from the $^{femur}T_{tibia}$ transformation matrix using Equation 7.10. The simpler Equation 7.6 can be used if the femur and tibia frames have parallel Z axes, as shown in Figure 7.11.

The unconstrained inverse kinematics approach is popular due to its ease of implementation and its availability in many commercial software packages. However, this approach has a major limitation: it does not account for the constraints imposed by anatomy. As a result, the model's body segments can change length, its joints can dislocate during movement, and the algorithm is sensitive to soft-tissue motions. Unconstrained inverse kinematic models produce errors that affect the computed joint angles and all subsequent calculations (e.g., joint moments and muscle forces). I prefer to use an alternative approach, which is described next.

Constrained inverse kinematics

The constrained inverse kinematics approach uses optimization to minimize the distance between the locations of experimental markers affixed to the subject and the locations of corresponding markers affixed to a kinematic model of the subject's skeleton (Figure 7.14). The model is composed of rigid bodies interconnected by anatomically accurate joints that permit only motions that can be reliably measured. The model is scaled to match the dimensions of the subject. Using a model that incorporates knowledge about the subject's geometry and mobility can produce more accurate estimates of joint angles than the unconstrained inverse kinematics approach. The constraints in such a model limit the set of possible solutions to those that represent anatomically feasible movements. For example, the lengths of the bones cannot change over time and the joints can neither dislocate nor move beyond defined anatomical limits. The following objective function $J(\underline{q})$

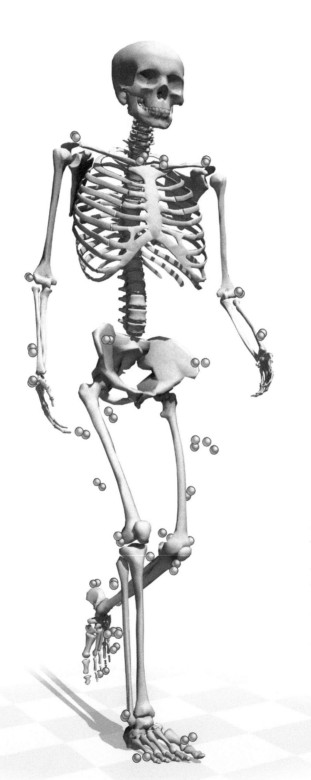

FIGURE 7.14
An inverse kinematics algorithm may produce more accurate estimates of joint angles if the solutions are constrained by an underlying skeletal model, where markers placed on the model (orange) track the trajectories of markers placed on the subject (blue).

is minimized over the model's generalized coordinates $\underline{q}$ (e.g., its location in space and the angles of each joint) at each point in time:

$$J(\underline{q}) = \sum_{k \in \text{Markers}} w_k \left\| \underline{x}_k^{\text{exp}} - \underline{x}_k(\underline{q}) \right\|^2 \qquad (7.16)$$

where $\underline{x}_k^{\text{exp}}$ is the location of the kth experimental marker in a global reference frame, $\underline{x}_k(\underline{q})$ is the location of the corresponding marker on the model (which is a function of the model's generalized coordinates), and w_k is the relative penalty associated with errors in the location of marker k. Ideally, each marker attached to the model would follow the same path through space as the corresponding experimental marker. The penalties encourage this result but do not demand it, allowing for a little bit of uncertainty in the data. A high weight w_k relative to the others puts a relatively high cost on any differences between the position of the kth experimental marker and the position of the corresponding model marker. You can think of each weight as the stiffness of a spring between the experimental marker and the marker in the model. The goal of the optimization is to minimize the energy stored in these imaginary springs. This goal can be posed as a weighted least-squares problem and solved using a suitable numerical algorithm, such as the Broyden–Fletcher–Goldfarb–Shanno (BFGS) iterative method (Nocedal and Wright, 2006).

Although the constrained inverse kinematics approach can produce more accurate results than the unconstrained approach, two key conditions must be satisfied for the method to be successful. First, the kinematic model must be properly scaled and constrained to represent the subject's geometry and the range of motion of each joint. For example, a two-dimensional model may suffice for computing the knee flexion angle during gait, but would be unsuitable for studying frontal-plane kinematics. The second condition is that the markers must be correctly registered—that is, they must be placed on the model to match the locations of experimental markers placed on the subject. Model scaling and marker registration can be time-consuming tasks, but well-designed software tools can alleviate this burden. My laboratory always computes joint angles using a constrained inverse kinematics approach because of its superior accuracy, especially for complex joints like the shoulder.

Kinematic model of the shoulder

The human shoulder is a complex mechanism comprising the scapula, clavicle, humerus, and the muscles and ligaments connecting them. A healthy shoulder provides a large range of motion and the structural support necessary to perform daily activities. The motion of the scapula is often used as a diagnostic indicator of shoulder pathologies like rotator cuff injuries. However, because the scapula moves underneath layers of muscle, fat, and skin, it can be difficult to determine its position and orientation accurately. Thus, it is difficult to identify patients with abnormal shoulder motions. A kinematic model can be used to refine our measurements by reducing the space of allowable joint motions. The model acts as a kinematic filter that removes infeasible motions and corrects for measurement noise.

Ajay Seth and colleagues developed a kinematic model of the shoulder that enables estimation of shoulder motion to a clinically relevant accuracy by tracking markers affixed to the skin. The model uses four degrees of freedom to define the position and orientation of the scapula relative to the thorax (Figure 7.15). The abduction–adduction and elevation–depression degrees of freedom are analogous to longitude and latitude, defining the position of the scapula as it glides over an ellipsoid that represents the surface of the thorax. The upward rotation degree of freedom allows the scapula to rotate about an axis normal to the thoracic surface. Finally, the internal rotation or "winging" degree of freedom enables the scapula to rotate about a longitudinal axis in the plane of the scapula (with the axis remaining tangent to the thoracic surface). This last degree of freedom allows the medial border of the scapula to lift off the surface of the thorax. The surface dimensions and joint axes can be customized to match an individual patient.

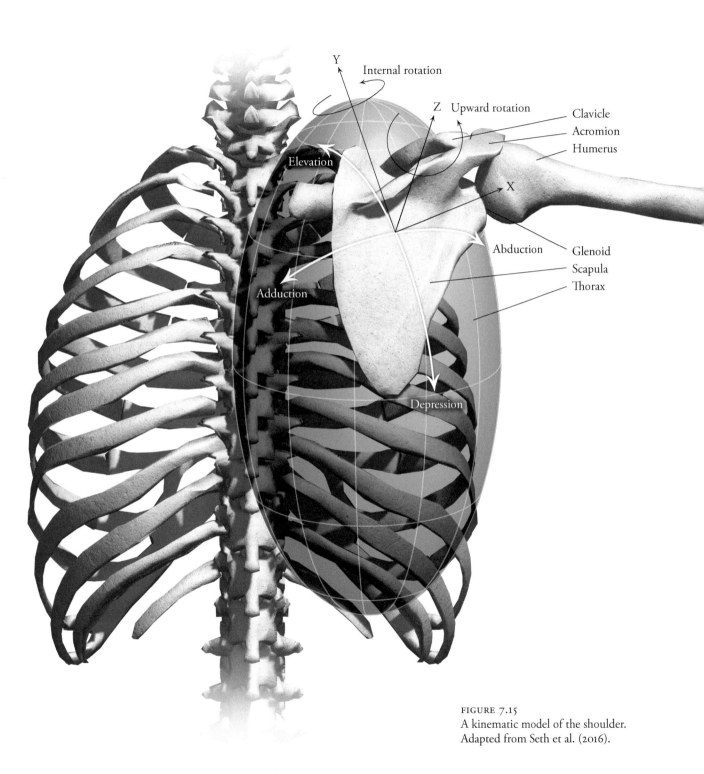

FIGURE 7.15
A kinematic model of the shoulder.
Adapted from Seth et al. (2016).

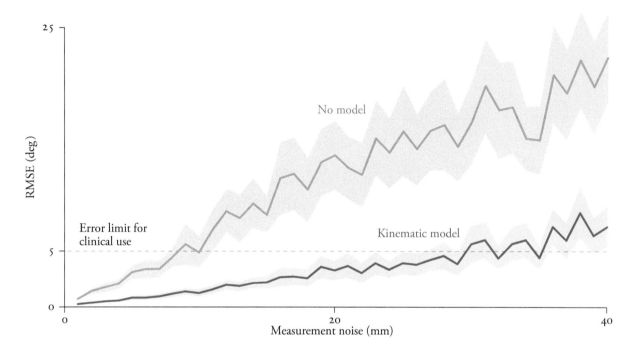

To validate their model, Ajay and colleagues compared the motion of the scapula in the model to the motion of bone-anchored pins that were tracked in the lab using a motion capture system (Figure 7.16). The shoulder model reduced root-mean-squared error in scapular kinematics due to skin-mounted marker noise by 65 percent compared to a solution computed without a model (i.e., using unconstrained inverse kinematics). This increase in accuracy allows clinical use of measurements with up to 20 mm of marker noise. Similar improvements in tracking accuracy can be obtained for gait studies using models of the lower limb.

FIGURE 7.16
Root-mean-squared error (RMSE) of scapular kinematics in the presence of measurement noise during a shoulder abduction task (mean ±1 standard deviation). Kinematic models attenuate noise due to soft-tissue motion by prohibiting nonphysiological motions, enabling clinical use. Adapted from Seth et al. (2016).

Assessing anterior cruciate ligament injury risk

The anterior cruciate ligament (ACL) is a key part of the knee's stabilization system. It connects the femur and tibia, and crosses in front of the posterior cruciate ligament (PCL). This "cross" shape gives the ligaments their name ("cruciate"). Roughly speaking, the ACL keeps the tibia from moving forward relative to the femur, while the PCL keeps it from moving backward. The PCL is rather hard to injure, but injuries to the ACL are staggeringly common and usually happen without any initiating contact. Over a quarter-million ACL repair surgeries are performed in the United States every year.

Female athletes are at much greater risk of ACL injury than male athletes, with one out of every 50–70 female athletes suffering this injury per year (Bates and Hewett, 2016). Several explanations have been proposed for this disparity, but a consistent finding of many biomechanics studies is that the ACL is placed under greatest stress when the knee is in a valgus position (Figure 7.17).

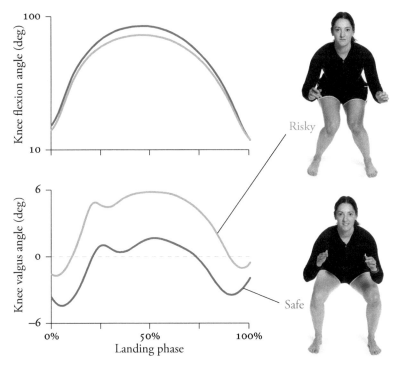

FIGURE 7.17
Soccer players landing in a risky pose where the knee is in a valgus position (top photo; orange curves) and in a safe pose (bottom photo; blue curves). The plots show the average knee flexion and valgus angles during landing, computed from motion capture and a constrained inverse kinematic model, averaged over 13 subjects. Adapted from Thompson et al. (2016).

The good news is that some ACL tears can be prevented. Both female and male athletes can be trained to avoid landing with the knee in a valgus alignment. We wanted to see whether we could help prevent ACL injuries and began by performing motion analysis on members of the Stanford women's soccer team. We found that these athletes are so strong and well trained that very few of them have biomechanical features that put them at risk for ACL injuries, at least during tasks we could perform in the laboratory. We then decided to study a population that had yet to be examined, preadolescent girls. My daughter's soccer coach helped us recruit 51 female athletes from local soccer teams. The athletes were 10 to 12 years old, about half of whom participated in training sessions to improve their biomechanics. The sessions were orchestrated by Julie Thompson, the leader of the study, who attended each team's practice sessions for 7–8 weeks. The training sessions were approximately 25 minutes long and replaced each team's standard warm-up before the start of regular practice. After the training program was complete, the girls came back into our lab for biomechanical analysis. The encouraging result is that this training program was effective at reducing a key risk factor for ACL injury, knee valgus alignment during a double-legged jump task.

Inverse kinematic modeling as we described above can also compute not just what an athlete is doing, but also what he or she should be doing. Cyril Donnelly recorded motion capture video of 34 Australian-rules football players performing a quick movement or "cut" to the side, which is common in sports and places the knee of the stance leg into the vulnerable valgus position (Donnelly et al., 2012). He identified the nine volunteers with the largest valgus moments and then used our OpenSim software to find anatomically feasible movements that would have placed the least stress on the ACL. OpenSim searched for poses of the model that satisfied two criteria: the pose must be a close match to the athlete's actual posture as recorded by the mocap system, and the pose must be less stressful for the ACL.

One strategy consistently worked for all nine volunteers: to move the body's center of mass slightly toward the desired direction of travel. Even a shift as small as 3 cm reduced the peak valgus moment by 40 percent. The benefit of the kinematic model is, first, that it identifies the center-of-mass shift as the most effective strategy to reduce the valgus moment during these maneuvers, and second, that it can be tailored to an individual athlete. In this way, an athlete can learn from the computer what will work best for him or her.

As you can see, subtle differences in the kinematics of different athletes can lead to big differences in injury risk. By quantifying movement and teasing out these differences, we have an opportunity to avoid unnecessary injuries. In this case, 3 cm of prevention may be worth $30,000 of cure!

8 Inverse Dynamics

To every action there is always opposed an equal reaction.
—Sir Isaac Newton

BY THE TIME MY MOTHER was 75 years old she could barely walk. Her right knee hurt with every step, and ascending stairs was out of the question. Too much pain. She wanted to play with my kids but could not keep up. Not being able to walk limited her participation in life and took a psychological toll. An X-ray of her knee showed that she had severe osteoarthritis, a degenerative joint disease that affects over 30 million individuals in the United States (Figure 8.1).

Osteoarthritis occurs when the cartilage surfaces that cap the ends of bones wear out. Cartilage is slippery and unequipped to sense pain, so it's an excellent material for allowing our joints to move freely without us feeling a thing. When cartilage wears out, however, it exposes the underlying bone, which is extremely sensitive to pressure and produces more friction, leading to joints that feel painful and stiff.

Overloading the knee contributes to the wear and tear that leads to osteoarthritis, yet it is impossible to measure the loads within a joint without performing a surgery to install a sensor in it. This has been done in only a few brave people who needed surgery for other reasons. Thus, we usually resort to other methods for estimating the loads within joints. Solving an inverse dynamics problem provides some of the information we need to estimate joint contact loads, an essential step toward understanding why biological joints fail.

When engineered products fail, forensic engineers seek to determine the root causes of these failures and, ultimately, the legal

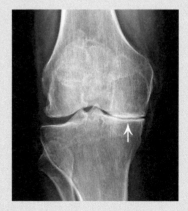

FIGURE 8.1
X-ray of a knee showing signs of osteoarthritis. Notice that the bones are in contact on the medial side of the knee (see arrow); the lateral side has retained more cartilage, which appears as space between the femur and tibia. Image courtesy of Julie Thompson-Kolesar.

liability of those involved. For example, forensic engineers with expertise in biomechanics are employed by law firms to evaluate the legitimacy of injury claims following vehicle collisions. Physical evidence like tire skid marks can be used to estimate the forces and accelerations during the collision. The process of reconstructing an accident is an example of an inverse problem because indirect observations are used to study phenomena for which one is unable to obtain direct measurements.

In Chapter 7, we introduced the *inverse kinematics* problem, an example of which is determining the pose of a subject's skeleton from measurements of the locations of skin-mounted optical markers in a global reference frame. Our goal in that chapter was limited; we were simply answering the question: What happened? In other words, what joint angles must have been present to produce the marker trajectories that were observed? In this chapter, we dig deeper and ask: What made it happen? Specifically, what forces and moments were responsible for producing the observed motion?

The process of determining forces and moments from kinematics is known as *inverse dynamics*. It is different from inverse kinematics because we are interested in the forces that produce the motion, not just the motion itself. It is also different from *forward dynamics*, where we predict the motion that will result when specified forces are applied to a system.

An inverse dynamics problem can be solved in a few ways. Typically, we use measurements of the ground reaction forces if they are available, so we will begin by briefly discussing techniques for measuring these forces. We then show a detailed example of how to compute joint moments during a squat. In this example, we work from the ground up, using measurements of the ground reaction forces and propagating this information upward through the leg joints. It is also possible to work from the top down, if we have reliable measurements of the motions of all the body segments.

Note that the forces we compute in this chapter are *not* the forces that would be measured in a biological joint. To compute joint contact loads, we must also estimate the forces applied by our muscles. This is a nontrivial problem and will be the focus of

Chapter 9. For now, just note that the same motion and external forces could be measured as a result of vastly different muscle coordination strategies, which would produce different joint contact loads (yet the same solution to an inverse dynamics problem).

Interestingly, if we have access to the ground reaction forces as well as the motions of all the body segments, then we obtain an overdetermined system of equations. In this case, we can use the additional information to minimize the consequences of measurement error, a strategy that echoes the constrained inverse kinematics algorithm described in Chapter 7. We close this chapter with an example that demonstrates how an inverse dynamic analysis has been used to evaluate a nonsurgical treatment for patients with knee osteoarthritis. This treatment could have helped my mother had I thought of it twenty years earlier.

Measuring external forces

Newton's three laws of motion are stated in his landmark book *Philosophiæ naturalis principia mathematica* (*Mathematical Principles of Natural Philosophy*), first published in 1687. Newton's second law is among the most important equations ever derived; Andrew Motte's 1729 translation from the original Latin states it as follows:

> Law II: The alteration of motion is ever proportional to the motive force impress'd; and is made in the direction of the right line in which that force is impress'd.

In other words, the rate of change of a particle's momentum is equal to and in the direction of an applied force. Newton's second law is now generally known as $\underline{F} = m\underline{a}$, where $\underline{F}$ is the vector sum of all forces applied to a body, m is the mass of the body, and $\underline{a}$ is the (linear) acceleration of its center of mass. This version follows from the original version, provided that the mass of the body does not change during the motion, an assumption that is reasonable in most applications. (A notable exception is rocketry.) Leonhard Euler later extended Newton's laws to apply to rotational motion. Euler's second law tells us that the rate of change of a body's angular momentum about its center of mass is equal to the sum of all moments acting about this point. (See Equation 8.6 below for a formal mathematical statement.)

If the mass and acceleration of each body segment are known, we can use Newton's and Euler's laws to determine the forces and moments applied to each body. Depending on the movement under investigation, there may be external forces acting on the subject, such as gravity, air resistance, or ground reaction forces. These external forces appear in the free-body diagram of the subject and must be either measured or estimated. In most cases of human movement, it is reasonable to neglect forces due to air resistance and to use a gravitational acceleration of $g = 9.81$ m/s^2 for computing weight. We usually prefer to measure ground reaction forces and other contact forces whenever possible to improve the accuracy of our calculations, as demonstrated below.

The first devices for measuring foot–ground contact forces were designed in the late 1800s by Étienne-Jules Marey and his student Gaston Carlet. By embedding pressure transducers into the sole of a shoe, Marey and Carlet were able to estimate the forces exerted between the foot and the ground during gait. The characteristic double-humped vertical component of the ground reaction force during walking was first reported in Carlet's Ph.D. thesis, which was published in 1872—the same year that Leland Stanford hired Eadweard Muybridge to determine whether a horse was ever completely airborne when trotting (see Figure 7.1). I find it remarkable that two of the most important quantitative tools for modern biomechanics were developed simultaneously, by two researchers (Carlet and Muybridge) who almost certainly had no idea of what the other was doing.

In modern experiments, ground reaction forces are measured using one or more *force plates*: flat, rigid platforms supported by load sensors (Figure 8.2). The load sensors contain strain gauges or piezoelectric crystals that convert small platform displacements into voltages. While force plates are somewhat reminiscent of an ordinary spring-based bathroom scale, there are major differences in precision, speed, and richness of the data. Because the ground reaction forces can change rapidly, particularly during foot

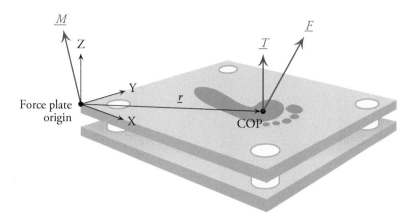

FIGURE 8.2
A force plate measures the forces and moments applied between the ground and the subject, collectively referred to as ground reaction forces. The center of pressure (COP) and the "free moment" ($\underline{T}$) can be computed from the resultant force ($\underline{F}$) and total moment ($\underline{M}$).

strikes, the data are typically collected at a rate of thousands of measurements per second. There is typically one load sensor located near each corner of a force plate, with each load sensor measuring strain and reporting force along three orthogonal axes. Thus, there are 12 independent force measurements, rather than just one as with a bathroom scale, which can be combined to compute the resultant force and moment relative to the force plate origin.

Center of pressure

Pressure is distributed unevenly over the bottom of the foot (Figure 8.3). This distributed pressure can be expressed as an equivalent resultant force applied at a point called the center of pressure. We can determine the center of pressure using the following moment-equivalence equation:

$$\underline{M} = \underline{r} \times \underline{F} + \underline{T} \qquad (8.1)$$

where $\underline{M}$ is the moment at the force plate origin, $\underline{r}$ is the vector from the origin to the center of pressure, $\underline{F}$ is the resultant ground reaction force, and $\underline{T}$ is the moment required to produce an equivalent system when the force $\underline{F}$ is applied at the center of pressure (Figure 8.2). In Equation 8.1, we have three independent equations (for M_x, M_y, and M_z) but six unknowns: the three components of vector $\underline{r}$ and the three components of moment $\underline{T}$. We obtain three additional equations by noting that the center of

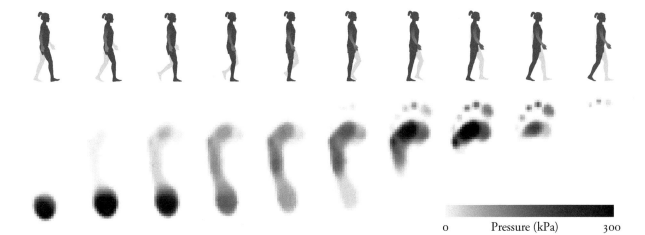

FIGURE 8.3
Pressure distributed over the foot during walking. Adapted from Pataky et al. (2012).

pressure remains on the surface of the force plate ($r_z = 0$) and that surface friction can generate a moment about only the vertical axis (T_z, which is also referred to as the "free moment"). While moments about the other two axes are conceivable, we would observe these moments only if we glued a person's foot to the force plate. It would then be possible for the sticky-footed subject to produce downward ground reaction forces and, thus, force couples about the other two axes.

In the absence of sticky feet, we can substitute $r_z = T_x = T_y = 0$ into Equation 8.1 and solve for the remaining unknowns. We obtain the following expressions for the location of the center of pressure relative to the force plate origin ($\underline{r}$) and for the free moment (T_z):

$$\underline{r} = \begin{Bmatrix} -M_y / F_z \\ M_x / F_z \\ 0 \end{Bmatrix} \tag{8.2}$$

$$T_z = M_z - r_x F_y + r_y F_x \tag{8.3}$$

It is useful to compute and visualize the center of pressure when verifying data-processing algorithms; a center of pressure located outside the footprint is an indication of calibration errors between the force plate and the motion capture system. It is also generally more intuitive to interpret forces and moments acting on the foot if they are transformed to the center of pressure. Note that the expressions in Equation 8.2 become ill-conditioned as the normal component of the ground reaction force (F_z) approaches zero, so the calculated center of pressure will be unreliable near foot contact and toe off.

Inverse dynamics algorithms

A dynamic analysis considers the motion of objects (kinematics) as well as the forces responsible for and generated by this motion (kinetics). In this section, we consider a typical inverse dynamics problem: *given* a representative model of a subject, the subject's joint kinematics over time, and measurements of the external forces applied to the subject, *find* the net joint forces and moments that must have been present to produce the given motion. In brief, we apply the laws of mechanics to each body segment in the model and compute the internal forces and moments acting at each joint. The joint angles are often estimated from optical marker trajectories using an inverse kinematics algorithm (Chapter 7), then smoothed and differentiated to obtain angular velocities and angular accelerations. It is possible to perform an inverse dynamic analysis without measurements of external forces, but we defer that discussion to later in this chapter. To demonstrate a simple inverse dynamics algorithm, we will compute the net joint forces and moments during the squat shown in Figure 8.4.

Before proceeding, it is worth considering the important question of how to select a suitable model for a particular study and, specifically, whether the simple model shown in Figure 8.4 is suitable for studying a squat. In general, the appropriate model fidelity—that is, the degree to which the model represents reality—depends on the research question being investigated. It is important to recognize that a more detailed model is not necessarily better. The objective when selecting a model is to maximize the model's utility, which often involves minimizing its complexity so that the simulation results are not polluted with irrelevant details. This

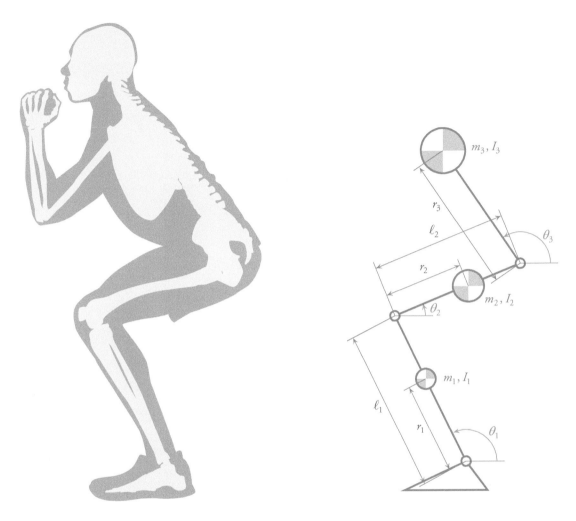

objective echoes the following wisdom from Einstein's 1934 paper "On the Method of Theoretical Physics":

> It can scarcely be denied that the supreme goal of all theory is to make the irreducible basic elements as simple and as few as possible without having to surrender the adequate representation of a single datum of experience.

Einstein was speaking of the relationship between theoretical and experimental physics, but the same sentiment holds here: we favor models that are minimally complex while being able to explain experimental observations and produce the desired outputs to a sufficient level of accuracy. Indeed, as famously stated by statisticians

FIGURE 8.4
Experimental setup (left) and approximate sagittal-plane model (right) for studying a squat. An inverse dynamic analysis computes the net forces and moments at each joint from the mass, inertia, geometry, and kinematics (positions, velocities, and accelerations) of each body segment (as well as external forces, when available).

George Box and Norman Draper, "all models are wrong, but some are useful."

Whether a particular model will be useful rather than just wrong depends on the context in which it is used. For example, it may be reasonable to model Mars as a point mass while simulating planetary orbits in our solar system or predicting the trajectory of an interplanetary spacecraft, but such a model would be insufficient for studying Martian weather patterns or simulating a spacecraft landing there. Conversely, using detailed planetary models in a study on orbital mechanics would increase modeling complexity and computational effort without adding value to the results. Back in the biomechanics world, we have noticed that some new users of our OpenSim software initially assume that they should always use a full-body model with dozens of degrees of freedom, when a much simpler model might tell them everything they need to know for their particular subject of interest. A single-degree-of-freedom knee joint model that can only flex and extend may be suitable for studying how knee extension moments vary with running speed, while a more detailed knee model with more degrees of freedom would be required for analyzing moments in the frontal plane in people with osteoarthritis.

Inverse dynamics with ground reaction forces

In this section, we will perform an inverse dynamic analysis of one instant of a two-legged squat using the planar model shown in Figure 8.4. The model consists of four rigid bodies connected by three revolute (pin) joints representing the ankle, knee, and hip. We assume that the joint angles, angular velocities, and angular accelerations are known from previously solving the inverse kinematics problem. We further assume that the subject's body is symmetrical and that any motion out of the sagittal plane can be neglected. We combine the left and right feet into a single rigid body ("the foot") that is assumed to have negligible mass and to remain stationary on the ground. The left and right shanks are also combined into a single rigid body, with mass and inertia that represent both shanks; the left and right thighs are similarly combined. We model the head, arms, and torso (HAT) as a single rigid body. We assume that each body segment of the model has been appropriately scaled to match the dimensions and mass properties of the corresponding body segments of the subject. In practice, model parameters such as lengths, masses, and inertias can be determined using a combination of direct measurements, published anthropometric tables, photographs, medical images, and other measurement and estimation techniques. The OpenSim models my group has developed include values for these parameters that can be scaled to represent a particular individual. Our models are freely available and can be accessed from the book's website.

In this simple example, we will begin at the feet and use the laws of motion, aided by a sequence of free-body diagrams, to compute the net forces and moments applied at each joint, starting with the ankle. For now, we assume that the ground reaction forces are known. The strategy of drawing free-body diagrams and performing hand calculations is sufficient for simple models like this one; in practice, we typically use software implementations of algorithms that are similar to the one we use below.

We have assumed that the dimensions and ground reaction forces are known, and that the feet are massless and stationary. Consequently, the only unknowns appearing in the free-body diagram of the foot segment (Figure 8.5) are the forces and moments applied at the ankle. Note that, by Newton's third law, forces and

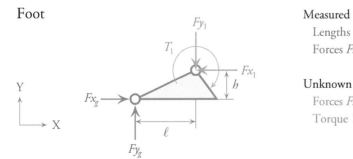

Foot

Measured
  Lengths ℓ, h
  Forces Fx_g, Fy_g

Unknown
  Forces Fx_1, Fy_1
  Torque T_1

FIGURE 8.5
Free-body diagram for the foot segment of the model shown in Figure 8.4.

moments that are of equal magnitude and opposite direction must be applied at the ankle on the shank body. We use the convention of drawing these vector pairs in the positive direction on the more proximal body and in the negative direction on the more distal body. For example, the Fx_1 vector points in the $-X$ direction in Figure 8.5, and the corresponding vector in the free-body diagram of the shank (Figure 8.6) will point in the $+X$ direction.

We can use Newton's second law and its rotational analogue to solve for Fx_1, Fy_1, and T_1. First, we sum the forces and apply Newton's second law in the X direction to compute Fx_1:

$$\sum F_X = m_0 \ddot{x}_0 \tag{8.4a}$$

$$Fx_g - Fx_1 = 0 \tag{8.4b}$$

$$Fx_1 = Fx_g \tag{8.4c}$$

where m_0 is the mass of the foot segment and $\ddot{x}_0$ is the horizontal acceleration of its center of mass, both of which we have assumed to be zero. Next, we sum the forces in the Y direction to compute Fy_1:

$$\sum F_Y = m_0 \ddot{y}_0 \tag{8.5a}$$

$$Fy_g - Fy_1 = 0 \tag{8.5b}$$

$$Fy_1 = Fy_g \tag{8.5c}$$

Finally, we solve for T_1. The rotational analogue of $\underline{F} = m\underline{a}$, which we mentioned above, has the following general form when moments are computed in a planar system about an arbitrary point P on a body:

$$\sum M_P = I\ddot{\theta} + \left(\underline{r}^P \times m\underline{a}\right) \cdot \hat{z} \tag{8.6}$$

where $\sum M_P$ is the sum of moments about point P, I is the moment of inertia of the body about its center of mass (COM), $\ddot{\theta}$ is the body's angular acceleration, $\underline{r}^P$ is the vector from point P to the COM, m is the mass of the body, and $\underline{a}$ is the linear acceleration of the body's COM. Because the product of mass and acceleration is force, we can interpret the second term on the right-hand side of Equation 8.6 as the projection of a moment onto the axis of rotation (as we saw in Equation 6.4).

Equivalent systems of equations are obtained regardless of the point about which moments are computed. For simplicity, we generally compute moments about the COM so that the cross product in Equation 8.6 is zero (specifically, so that $\underline{r}^P = \underline{0}$). However, we have assumed that the feet are massless and stationary. Thus, the right-hand side of Equation 8.6 is always zero and we can assume any location for the COM. For convenience, we sum moments about the ankle:

$$\sum M_A = 0 \tag{8.7a}$$

$$Fx_g h - Fy_g \ell - T_1 = 0 \tag{8.7b}$$

$$T_1 = Fx_g h - Fy_g \ell \tag{8.7c}$$

Equations 8.4c, 8.5c, and 8.7c comprise a linear system of equations that can be used to compute Fx_1, Fy_1, and T_1 given ℓ, h, Fx_g, and Fy_g.

We now repeat the process for the shank segment (Figure 8.6), where the forces and moments applied at the ankle (Fx_1, Fy_1, and T_1) are assumed to be known from the previous step and those at the knee (Fx_2, Fy_2, and T_2) are now the unknowns. We first compute the acceleration of the COM ($\ddot{x}_1$ and $\ddot{y}_1$) from the kinematics of the ankle joint (θ_1, $\dot{\theta}_1$, and $\ddot{\theta}_1$), noting that the ankle joint center is stationary:

$$x_1 = r_1 c\theta_1 \tag{8.8a}$$

$$\dot{x}_1 = -r_1 s\theta_1 \dot{\theta}_1 \tag{8.8b}$$

$$\ddot{x}_1 = -r_1 \left(s\theta_1 \ddot{\theta}_1 + c\theta_1 \dot{\theta}_1^2 \right) \tag{8.8c}$$

$$y_1 = r_1 s\theta_1 \tag{8.9a}$$

$$\dot{y}_1 = r_1 c\theta_1 \dot{\theta}_1 \tag{8.9b}$$

$$\ddot{y}_1 = r_1 \left(c\theta_1 \ddot{\theta}_1 - s\theta_1 \dot{\theta}_1^2 \right) \tag{8.9c}$$

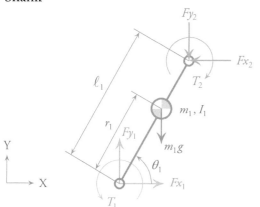

Shank

Measured
Lengths ℓ_1, r_1
Orientation θ_1
Mass m_1
Inertia I_1

Already computed
Forces Fx_1, Fy_1
Torque T_1

Unknown
Forces Fx_2, Fy_2
Torque T_2

FIGURE 8.6
Free-body diagram for the shank segment of the model shown in Figure 8.4. The pose of the body segment in a free-body diagram need not match its pose during the activity being studied. Although irrelevant mathematically, we draw the shank segment in this convenient canonical pose, where θ_1 is between 0 and 90 degrees.

where we have used $s\theta_1 \triangleq \sin(\theta_1)$ and $c\theta_1 \triangleq \cos(\theta_1)$ for notational convenience. We now apply Newton's second law in the X direction to compute Fx_2:

$$\sum F_X = m_1 \ddot{x}_1 \tag{8.10a}$$

$$Fx_1 - Fx_2 = -m_1 r_1 \left(s\theta_1 \ddot{\theta}_1 + c\theta_1 \dot{\theta}_1^2 \right) \tag{8.10b}$$

and again in the Y direction to compute Fy_2:

$$Fy_1 - Fy_2 - m_1 g = m_1 r_1 \left(c\theta_1 \ddot{\theta}_1 - s\theta_1 \dot{\theta}_1^2 \right) \tag{8.11}$$

Finally, we sum moments about the COM to obtain an expression for T_2:

$$\begin{aligned} T_1 - T_2 + Fx_1 \, r_1 \, s\theta_1 - Fy_1 \, r_1 \, c\theta_1 \\ + Fx_2 \, d_1 \, s\theta_1 - Fy_2 \, d_1 \, c\theta_1 = I_1 \ddot{\theta}_1 \end{aligned} \tag{8.12}$$

where $d_1 \triangleq \ell_1 - r_1$. Equations 8.10b, 8.11, and 8.12 comprise a second linear system of three equations in three unknowns, and we can now solve for the net forces and moments applied at both the ankle and the knee.

The same procedure is repeated for the thigh segment (Figure 8.7), where the unknowns are the forces and moments applied at the hip (Fx_3, Fy_3, and T_3). To compute the acceleration of the COM

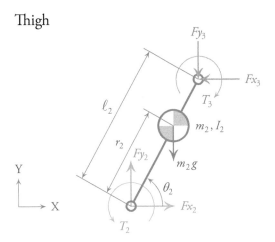

FIGURE 8.7
Free-body diagram for the thigh segment of the model shown in Figure 8.4.

Measured
Lengths ℓ_2, r_2
Orientation θ_2
Mass m_2
Inertia I_2

Already computed
Forces Fx_2, Fy_2
Torque T_2

Unknown
Forces Fx_3, Fy_3
Torque T_3

relative to ground ($\ddot{x}_2$ and $\ddot{y}_2$), note that the position of the hip relative to the ankle is a function of both θ_1 and θ_2 (see Figure 8.4):

$$x_2 = \ell_1 c\theta_1 + r_2 c\theta_2 \tag{8.13a}$$

$$\dot{x}_2 = -\ell_1 s\theta_1 \dot{\theta}_1 - r_2 s\theta_2 \dot{\theta}_2 \tag{8.13b}$$

$$\ddot{x}_2 = -\ell_1 \left(s\theta_1 \ddot{\theta}_1 + c\theta_1 \dot{\theta}_1^2\right) - r_2 \left(s\theta_2 \ddot{\theta}_2 + c\theta_2 \dot{\theta}_2^2\right) \tag{8.13c}$$

$$y_2 = \ell_1 s\theta_1 + r_2 s\theta_2 \tag{8.14a}$$

$$\dot{y}_2 = \ell_1 c\theta_1 \dot{\theta}_1 + r_2 c\theta_2 \dot{\theta}_2 \tag{8.14b}$$

$$\ddot{y}_2 = \ell_1 \left(c\theta_1 \ddot{\theta}_1 - s\theta_1 \dot{\theta}_1^2\right) + r_2 \left(c\theta_2 \ddot{\theta}_2 - s\theta_2 \dot{\theta}_2^2\right) \tag{8.14c}$$

The dynamic equations for the thigh segment are analogous to those we obtained for the shank. Notice that the coordinates of the knee are ($\ell_1 c\theta_1, \ell_1 s\theta_1$); these terms in Equations 8.13a and 8.14a could be replaced with measurements if the knee location were tracked directly.

In summary, we obtain a linear system of nine equations in nine unknowns: the net forces and moments applied at the ankle, knee, and hip. Thus, if the ground reaction forces are known, it is necessary to measure or otherwise obtain the geometric properties, inertial properties, and kinematics of only the foot, shank, and thigh

Head, arms, torso
(HAT)

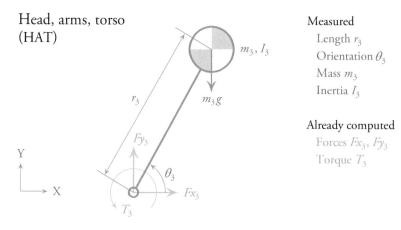

Measured
Length r_3
Orientation θ_3
Mass m_3
Inertia I_3

Already computed
Forces Fx_3, Fy_3
Torque T_3

FIGURE 8.8
Free-body diagram for the head, arms, and torso (HAT) of the model shown in Figure 8.4.

segments. Measurements of the motions of the head, arms, and torso are not needed, which can be an advantage because measurements of these body segments can be inaccurate. A free-body diagram of the HAT segment (Figure 8.8) confirms that all variables have now been computed, and thus analysis of the HAT is not needed when ground reaction forces are measured.

Inverse dynamics without ground reaction forces

If the ground reaction forces are not known, then there are five unknowns in the dynamic equations for the foot: Fx_1, Fy_1, T_1, Fx_g, and Fy_g. The system is underdetermined, and we require additional information to complete an inverse dynamic analysis. One strategy is to replace Equations 8.4c, 8.5c, and 8.7c with the dynamic equations for the HAT (Figure 8.8):

$$Fx_3 = m_3 \ddot{x}_3 \qquad (8.15a)$$

$$Fy_3 = m_3 \ddot{y}_3 + m_3 g \qquad (8.15b)$$

$$Fx_3\, r_3\, s\theta_3 - Fy_3\, r_3\, c\theta_3 + T_3 = I_3 \ddot{\theta}_3 \qquad (8.15c)$$

Provided that the kinematics of the HAT have been measured, Equations 8.15 comprise a linear system of three equations in three unknowns (Fx_3, Fy_3, and T_3). If we use Equations 8.15 instead of the three equations we derived from the free-body diagram of the foot segment, we obtain a system of nine equations in nine unknowns as we had in the previous section. Note, however, that our sequential calculations now begin at the HAT (sometimes referred to as a "top-down" algorithm) rather than the foot (a "bottom-up" algorithm). Thus, we would be relying on our measurements of the HAT segment kinematics (θ_3 and potentially noisy derivatives thereof) rather than measurements of the ground reaction forces (Fx_g and Fy_g). Recall our assumption that the head, arms, and torso act as a single rigid body whose inertia is constant and whose center of mass location is known. These assumptions are likely to introduce greater uncertainty in our results than measurements of the ground reaction forces, which are typically reliable as long as the force plate has been properly calibrated.

Verifying dynamic consistency

What if we have confidence in our measurements of both the ground reaction forces and the HAT segment kinematics? In that case, we have an overdetermined system (i.e., more equations than unknowns) and can use the "extra" information to improve the model performance. For example, we can begin at the foot segment and use the equations we derived when assuming that the ground reaction forces are known to obtain the net forces and moments applied at the hip (Fx_3, Fy_3, and T_3). We can then *apply* these forces to the HAT segment and use the model to predict the HAT segment's linear and angular accelerations (forward dynamics):

$$\tilde{\ddot{x}}_3 = \frac{1}{m_3} Fx_3 \tag{8.16a}$$

$$\tilde{\ddot{y}}_3 = \frac{1}{m_3} Fy_3 - g \tag{8.16b}$$

$$\tilde{\ddot{\theta}}_3 = \frac{1}{I_3} \left(Fx_3\, r_3\, s\theta_3 - Fy_3\, r_3\, c\theta_3 + T_3 \right) \tag{8.16c}$$

These predictions can then be compared with the measured accelerations of the torso to adjust model parameters such as the mass, inertia, and center of mass location of the HAT segment, parameters with which there is typically a high amount of uncertainty. A similar approach is to apply fictitious forces to the HAT segment's center of mass so that the model experiences the same accelerations that were measured (i.e., so that the model's motion is dynamically consistent with the observations), and then adjust the parameters of the model to minimize these fictitious or "residual" forces.

Adjusting model parameters to improve dynamic consistency can be a time-consuming and subjective task when done by hand. Art Kuo proposed a method to find the best compromise among all observations in an objective way, simply by assembling all dynamic equations into an overdetermined system of the form $A\underline{x} = \underline{b}$ and computing the pseudoinverse A^+ of matrix A. The solution $\underline{x} = A^+\underline{b}$ then represents a least-squares "best fit" to the measured ground reaction forces and joint kinematics (Kuo, 1998).

Now that we have seen how to solve a simple two-dimensional inverse dynamics problem by hand, we can move on to solve more realistic three-dimensional inverse dynamics problems. These tools allow us to estimate the net moments about the hip, knee, and ankle during walking and running.

Joint moments during walking and running

Edith Arnold and Sam Hamner were working in my lab and wanted to examine the changes in joint moments during walking and running at different speeds. They collected experimental marker positions and ground reaction forces as subjects walked and ran on a treadmill, and then used OpenSim to perform three-dimensional inverse kinematic and inverse dynamic analyses.

During walking and running, we see that the produced joint moments are generally greater during stance than during swing (Figures 8.9 and 8.10). Of all lower-limb joint moments, the peak ankle plantarflexion moment is greatest. At slow walking speeds, swing is largely passive (i.e., moments are low), an assertion that we will confirm when we investigate muscle activity during walking in Chapter 11. Also notice the systematic increases in joint moments with increasing walking and running speed, and that the joint

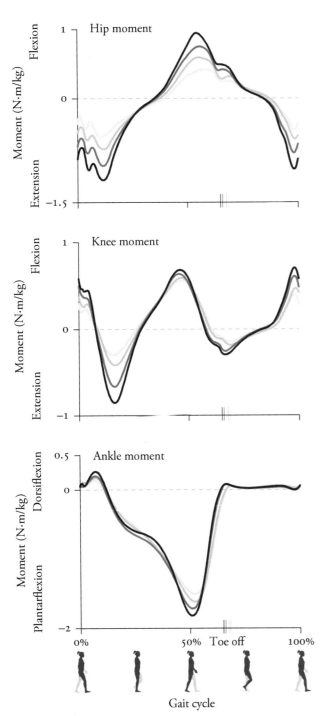
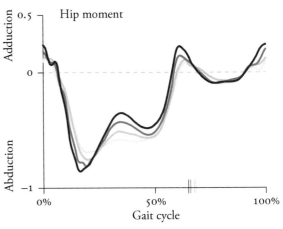

FIGURE 8.9
Representative joint moments over the gait cycle when walking at several speeds. Normalized by body mass and averaged over 10 subjects. Vertical lines on the horizontal axis indicate toe off at each speed. Data from Arnold et al. (2013).

INVERSE DYNAMICS 211

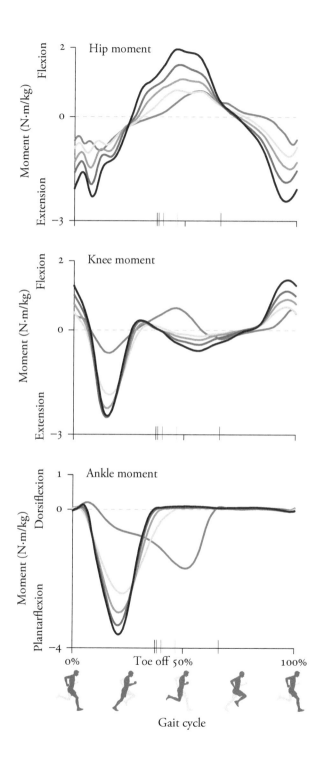

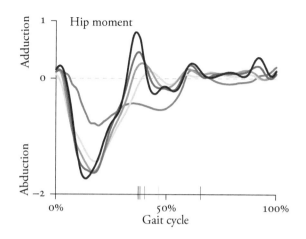

— 1.5 m/s (walking)
— 2 m/s
— 3 m/s
— 4 m/s
— 5 m/s

FIGURE 8.10
Representative joint moments over the gait cycle when running at several speeds (walking at 1.5 m/s from Figure 8.9 is shown for reference). Normalized by body mass and averaged over 10 subjects. Vertical lines on the horizontal axis indicate toe off at each speed. Data from Hamner and Delp (2013).

moments are higher in running than walking. These higher joint moments indicate greater muscle activity and, together with the higher ground reaction forces, result in larger joint loads during running than walking.

Gait retraining to reduce knee loads and pain

Knee osteoarthritis affects about 20 percent of adults over the age of 45 and can cause pain, restrict physical activity, and decrease one's quality of life. Although it is often treated by surgical intervention, such as a total joint replacement, it would be preferable to avoid surgery—provided that we can find another effective treatment.

Remarkably, Pete Shull and his colleagues have discovered a nonsurgical approach to treating knee osteoarthritis that reduces knee loads and relieves pain. In many patients, an effective treatment is simply to learn how to walk with a reduced foot progression angle, or with more of a toe-in gait (Figure 8.11, top right).

We can understand why this strategy works by looking at the geometry of knee joints and the frontal-plane dynamics during walking. Throughout most of this book, we have focused on motions and moments in the sagittal plane; however, the motions and moments in the frontal plane are critical in the development and treatment of knee osteoarthritis. As shown in Figure 8.11, the ground reaction force generates an external knee adduction moment during stance. (Please note that our convention in this book is to report the "internal" moments generated by muscles, as opposed to the "external" moments of ground reaction forces; but we report external moments in this example to match the literature on this topic.) This moment acts to load the medial compartment of the knee, which typically supports about 50 percent more weight than the lateral compartment during walking. The peak knee adduction moment typically occurs in early stance; in individuals with knee osteoarthritis, the magnitude of this peak has been linked to the severity and progression of

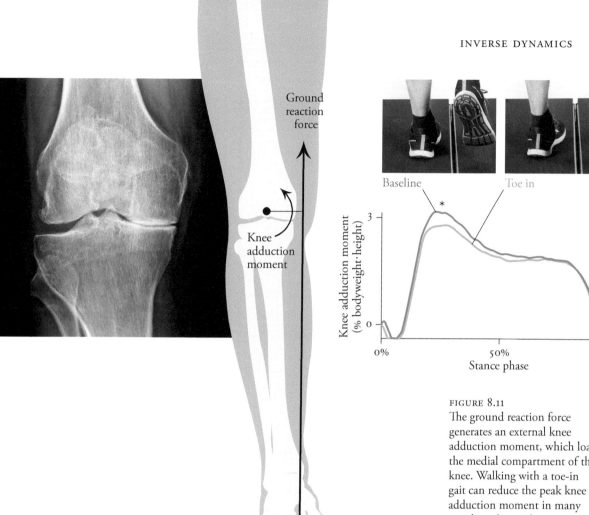

FIGURE 8.11
The ground reaction force generates an external knee adduction moment, which loads the medial compartment of the knee. Walking with a toe-in gait can reduce the peak knee adduction moment in many people and provides a nonsurgical treatment option for patients with medial compartment knee osteoarthritis. Data from Shull et al. (2013). X-ray image courtesy of Julie Thompson-Kolesar.

medial compartment osteoarthritis. Note that a large ground reaction force (e.g., due to obesity or engaging in high-impact activities) increases the load on the knee and can contribute to a loss of cartilage in the medial compartment. A lack of exercise can also lead to poor cartilage health.

Once osteoarthritis has developed, reducing the knee adduction moment (and therefore the load on the medial compartment) can alleviate pain and improve function. Surgery can reduce the knee adduction moment by up to 30 percent. But as shown in Figure 8.11, patients with medial compartment osteoarthritis were able to reduce the adduction moment simply by adjusting their foot progression angle (i.e., walking with more of a toe-in gait). This improvement reduced the adduction moment more than other nonsurgical treatments, such as braces or insoles. Decreasing the foot progression angle shifted the knee joint center medially and the center of pressure laterally, both of which reduced the moment arm of the ground reaction force at the knee. This simple gait modification was easy to learn and remarkably effective, decreasing pain and improving function in the study participants (Shull et al., 2013). Scott Uhlrich has recently shown that even greater improvements can be achieved by personalizing the gait retraining program (Uhlrich et al., 2018).

In this chapter, we have seen that inverse dynamics is a useful technique for computing the net forces and moments that must be present at each joint to produce an observed motion. However, we have not yet discussed the role of muscles, so our inverse dynamics story is still incomplete. This omission is important for two reasons. First, as we mentioned at the outset, the net joint forces computed in this chapter are not the same as the contact forces experienced by the cartilage in the joint. For example, we can easily imagine two ways in which a net joint moment can be zero. In one scenario, the muscles spanning the joint are inactive and generating no force. In another scenario, two muscles on opposite sides of the joint are generating large forces but producing equal and opposite moments about the joint. In each case, the net joint moment is zero, but the joint contact forces in the second scenario would be higher. We would expect to see more wear and tear on the cartilage in the second scenario because of the high compressive loads produced by the muscles, even though the net joint moment is the same.

A second reason to take account of individual muscle forces is that our bodies have a great deal of redundancy. There are many ways to accomplish most of the tasks we ask our muscles to do. That's a good thing: it means that if one muscle is fatigued or injured, others can pick up the slack. Net forces tell us what the body *must* do, but by bringing individual muscles into our models, we start to appreciate the variety of options for accomplishing these tasks. Understanding how our brain chooses among the available options is one of the most fascinating problems in biomechanics, a problem we will turn to in the next chapter.

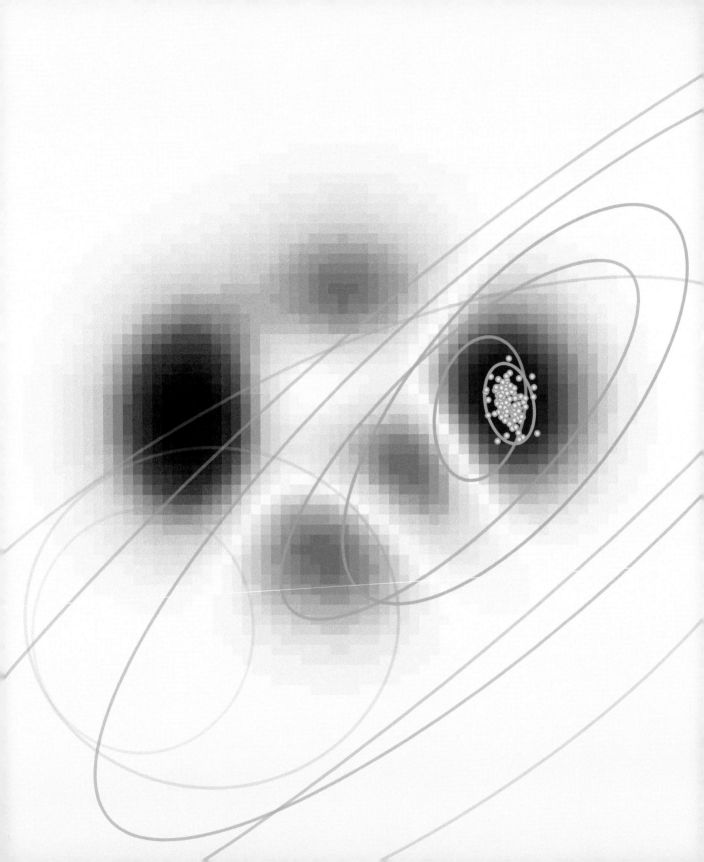

9 Muscle Force Optimization

There is always a well-known solution to every human problem—neat, plausible, and wrong.
—H. L. Mencken

ONE OF THE GREAT biomechanics discoveries of my lifetime did not occur in a laboratory, and did not involve any special equipment other than the marvelously adaptable human body. In 1968, a 21-year-old civil engineering student named Dick Fosbury revealed his discovery and astounded the sports world by going to the Olympics and jumping over the high-jump bar *backward*. "Fosbury goes over the bar like a guy being pushed out of a 30th-story window," wrote one sportswriter. Yet this seemingly ungainly style took him to an Olympic record height of 2.24 meters and a gold medal.

Fosbury actually began using his signature technique in high school, out of desperation. He had failed to master the standard technique of the time, the "Western roll," in which the jumper crosses the bar face down, as if embracing it with his arms and legs. He also tried the older "scissors" technique, in which the jumper clears the bar in almost a seated posture while moving his legs up and down like scissors. But this technique is less than ideal because the jumper has to propel his center of mass far above the bar.

In a Fosbury flop, the jumper runs toward the bar at one end, curves inward toward the middle, and then twists at the last moment so that he or she jumps over the bar backward (Figure 9.1). One part of the body slithers over the bar at a time: first the head and shoulders, then the torso, the hips, the knees, and finally the feet. An advantage of the Fosbury flop is that the body's center of

FIGURE 9.1
The "Fosbury flop" high-jump technique. Photo of Ma'ayan Furman-Shahaf by Evren Kalinbacak.

mass never has to go over the bar. At peak altitude, when the hips are above the bar, the back is arched so that the head and legs dangle below the bar, lowering the center of mass.

Although Fosbury discovered the flop without any formal biomechanics training, we have now caught up to his pioneering genius. Coaches now use high-speed video to instruct their high jumpers on where to plant their foot, how low to crouch before pushing off the ground, and so on. They have thought of improvements Fosbury never dreamed of. Lifting the arms at takeoff increases the vertical kinetic energy. Tucking the chin after the hips clear the bar helps to force the hips down and, by Newton's third law, lifts the feet up and over the bar.

Since 1980, every high-jump world record has been set using the Fosbury flop, and all other techniques have disappeared from the world stage. Now the debate is over which kind of Fosbury flop is best: the "speed flop" or the "power flop"?

For me, the Fosbury flop story teaches several valuable lessons. The first is that we should never assume that the popular way to do something is the only way or the best way. But a less philosophical and more biomechanical message is that our body, with an articulating skeleton and hundreds of muscles, gives us an infinite variety of ways to accomplish a task, whether it's an ostensibly simple one like reaching for a cup of coffee or a complex one like high-jumping two meters. Every action requires our brain to coordinate the activity of our muscles. A simple motion of our chin can have effects all the way down to our feet.

To perform any motion, our bodies solve an optimization problem, trying to get the maximum result out of a minimum effort. But one catch, which can trip up both mathematical models and real-world athletes, is that a *local* optimum is not always the *global* optimum. High-jumpers before Fosbury thought they had found the best technique. They were wrong. But they could not raise the bar merely by making small ("local") adjustments to the Western roll technique. They needed an athlete-engineer who was willing to make large ("global") adjustments that would change our idea of what a high jump should look like.

In this chapter, we explore two broad scenarios in which numerical optimization is used to compute muscle forces. The first scenario arises when we wish to estimate the muscle forces that were responsible for generating an observed motion. Because of the redundancy of the human musculoskeletal system, there are many possible solutions in general, and we will apply optimization to find the set of muscle forces that is the best by some criterion. We call this problem the *muscle redundancy problem* (it is also known as the muscle force-sharing problem). The second broad scenario arises when studying the muscle coordination that optimizes performance of a given task, such as the high jump. In this scenario, we need to compute muscle forces as well as the body segmental motions that achieve the task. If an optimizer were provided with the rules of the high-jump competition and a sufficiently realistic model of the musculoskeletal system, it would discover the Fosbury flop (as well as other techniques we may never have seen before). Note that estimating muscle forces using optimization is not merely a computational convenience: the nervous system is, itself, an optimizer of sorts. Let's jump into the details here.

Biological and numerical optimizers

People consciously optimize their everyday lives. We make decisions about how to allocate our time, money, attention, and other limited resources to maximize some balance of short-term and long-term satisfaction, and do so while adhering to constraints like laws and other obligations. Amazingly, humans optimize subconsciously as well. Recall from Chapter 2 that we naturally adjust our walking speed, cadence, and other variables of gait to keep our cost of transport near a minimum—unless we're in a hurry, of course, in which case we prioritize speed over metabolic cost. We also optimize when learning new tasks and adapting to new situations, maximizing some combination of accuracy, efficiency, comfort, and other performance metrics depending on the situation. Experiments have shown that we maintain an "internal model" of our bodies, a subconscious understanding of the relationship between the amount we excite our muscles and the resulting motions of our body segments. When exposed to a new environment, such as microgravity aboard the International Space Station or an artificial force field imposed by a robotic device, we are initially clumsy and inefficient because we are planning and anticipating our movements using an internal model that is no longer accurate. As we explore the new environment, we use visual and other sensory feedback to adjust our internal model and improve our coordination. Over time, we adapt to the new environment and regain proficiency by relearning the relationship between muscle excitations and performance.

Numerical optimization algorithms use similar exploratory or "guess-and-check" strategies to find solutions to underdetermined problems. We refer to the guesses as *candidate solutions*, each of which proposes a numerical value for all unknowns or *design variables*, such as muscle excitations in the muscle redundancy problem. The suitability of each candidate solution is determined by evaluating the *objective function* or *cost function*, an expression that quantifies the favorability of a particular set of design-variable values. The candidate solution that renders the best objective function value (minimum or maximum, as the case may be) is referred to as the *optimal solution*, or simply the solution to the optimization problem. (Note that we may define our task as always seeking to *minimize* the objective function; to *maximize* a quantity like jump height, we simply minimize its negative.) As we will see, the choice of objective function can have a profound effect on the solution and the amount of effort required to compute it.

The optimal solution is also influenced by *constraints*. In many problems, we demand that certain expressions (equalities or inequalities) are satisfied by a candidate solution for it to be *feasible*, or acceptable as a solution. In the muscle redundancy problem, for example, we require that all muscle forces be tensile and that the moments they generate sum to the desired net joint moments (computed using an inverse dynamics algorithm, for example). Because constraints can only reduce the number of feasible solutions (i.e., they decrease the size of the *solution space*), the solution to a constrained optimization problem has an objective function value that is no better than if the optimization problem were unconstrained. Adding constraints reduces your options. If there are too many constraints, the solution space may be empty (the problem may have no feasible solutions). Depending on the research question, it may be acceptable to convert one or more strict constraints into soft constraints, which take the form of "penalties" that are added to the objective function. Penalty terms are typically scaled by weighting factors so that their contributions to the objective function can be adjusted according to their relative importance. A generic optimization problem may be stated as follows:

Generic Optimization Problem

$$\text{minimize} \quad J(\underline{x})$$

Adjust design variables $\underline{x}$ to minimize objective function $J(\underline{x})$

$$\text{subject to} \quad g_i(\underline{x}) \leq 0, \quad i = 1, \ldots, n^i$$

while satisfying n^i inequality constraints,

$$h_j(\underline{x}) = 0, \quad j = 1, \ldots, n^j$$

satisfying n^j equality constraints,

$$\underline{x}^{\text{lower}} \leq \underline{x} \leq \underline{x}^{\text{upper}}$$

and respecting bounds on the design variables.

We subconsciously coordinate our muscles using these same principles. Consider the action of picking up a pen that is sitting on your desk. You know the required final position and orientation of your hand, but you are free (within limits) to choose the final pose of your arm and the path to arrive at this pose. Stated formally, each candidate solution to this problem might propose a different set of muscle excitations defined as parameterized functions of time; these parameters would be the design variables. Of all the candidate solutions, the feasible solutions would be those describing muscle excitation patterns that achieve the desired final hand pose. Additional constraints might describe other nonnegotiable requirements, such as avoiding obstacles. You would presumably seek to minimize travel time to some degree, but the objective function would also likely include a term that favors low muscle effort for this nonurgent task. A naive approach to solving such problems is to evaluate the objective function for all feasible solutions and select the best candidate. As you might expect, however, a so-called "exhaustive search" (in which every possibility is considered) is inefficient and impractical for all but the simplest problems. A numerical optimizer attempts to find the most favorable feasible solution (or at least a feasible solution that is good enough for the problem at hand) by evaluating only as many candidate solutions as it deems necessary. Many optimizers differ only in how they select new candidate solutions to evaluate based on the suitability of those that have already been explored.

Our exploration now moves to discussing strategies for solving the muscle redundancy problem. For simplicity, we will seek a solution at a single instant of time, assuming that the required net joint moments are known and that the subject is stationary. We sometimes refer to such problems as "static" optimization problems because some of the time-dependent aspects of the problem are ignored. In the next section, we formulate an optimization problem statement and solve a simple muscle redundancy problem by inspection to build intuition about the solution space. In the subsequent sections, we discuss numerical algorithms that can solve the nontrivial optimization problems that we encounter in the real world.

Static optimization problems solved by inspection

We will first focus on the model shown in Figure 9.2. Suppose we wish to generate a net ankle plantarflexion moment of 100 N·m. Which muscles should be excited to generate this moment? There are many feasible solutions. For example, perhaps the entire desired moment should be generated by the soleus, or perhaps it would be best if each of the three plantarflexors in the model contributed some fraction of the total. Forces could be generated by one or both of the dorsiflexor muscles as well. As given, the problem is underdetermined: there are fewer equations than unknowns and we need more information to solve the problem. Stated formally, we wish to find the force F_i generated by each muscle i (our design variables) such that the total ankle plantarflexion moment is 100 N·m (an equality constraint), subject to the condition that F_i is positive and does not exceed F_i^{max} for any $i = 1, \ldots, 5$ (bounds).

One approach to solving an underdetermined problem is to modify the problem so that there are as many equations as unknowns. In our example, we will assume for simplicity that the dorsiflexors are inactive and generate zero force. This assumption reduces the number of unknowns from five to three: the force generated by the gastrocnemius (which we will call F^{GAS}), soleus

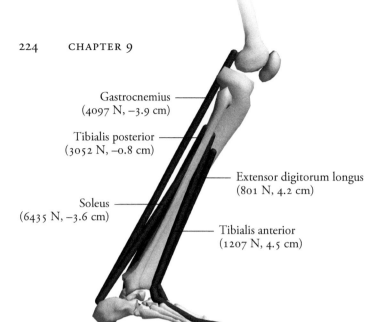

FIGURE 9.2
A musculoskeletal model of the shank and foot with the key plantarflexor and dorsiflexor muscles. A measured ankle moment could have been generated by many combinations of muscle forces. Values in parentheses are the instantaneous maximum force (assuming zero velocity and a rigid tendon) and moment arm at the ankle corresponding to the pose shown.

Gastrocnemius (4097 N, −3.9 cm)
Tibialis posterior (3052 N, −0.8 cm)
Extensor digitorum longus (801 N, 4.2 cm)
Soleus (6435 N, −3.6 cm)
Tibialis anterior (1207 N, 4.5 cm)

(F^{SOL}), and tibialis posterior (F^{TP}). However, we still have only one equation (the equality constraint specifying the desired net ankle joint moment). We could assume that each plantarflexor muscle will generate the same force and introduce two additional equations, $F^{SOL} = F^{GAS}$ and $F^{TP} = F^{GAS}$. This strategy will lead to a feasible solution, but will not produce physiologically reasonable results in general. Instead, we will define an objective function, $J(\underline{F})$, that quantifies the favorability of a solution as a function of muscle forces $\underline{F}$, and will then solve the following optimization problem:

Optimization Problem 1: Find the ankle plantarflexor muscle forces that produce a desired joint moment

minimize $\quad J(\underline{F})$ — Objective function where smaller values are favored

subject to $\quad 0.039 F^{GAS} + 0.036 F^{SOL} + 0.008 F^{TP} = 100$ — Muscles must produce the desired net ankle moment

$$0 \leq F^{GAS} \leq 4097$$
$$0 \leq F^{SOL} \leq 6435$$
$$0 \leq F^{TP} \leq 3052$$

— Muscle forces must lie within physiological ranges

For example, suppose we assume that the nervous system minimizes the total muscle force generated at each instant of time, so that $J(\underline{F})$ is the sum of the instantaneous muscle forces:

$$J(\underline{F}) \triangleq F^{\text{GAS}} + F^{\text{SOL}} + F^{\text{TP}} = \sum_{i=1}^{3} F_i \qquad (9.1)$$

The solution in this case is $F^{\text{GAS}} = 2564$, $F^{\text{SOL}} = 0$, and $F^{\text{TP}} = 0$, because the gastrocnemius has the largest moment arm and therefore can generate the desired ankle plantarflexion moment using the least amount of force. Note that the gastrocnemius is capable of generating the entire 100 N·m of plantarflexion moment in this model (i.e., 2564 N < 4097 N), and in fact it can generate a moment of about 160 N·m in this pose when fully activated. Ankle moments in excess of 160 N·m would be generated by recruiting the soleus as well (the muscle with the second-largest moment arm), followed by the tibialis posterior once the soleus had also reached its maximum force (Figure 9.3).

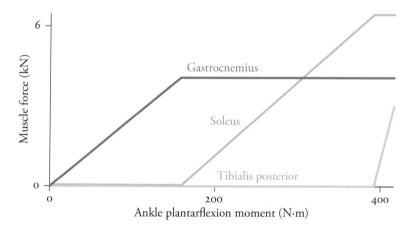

FIGURE 9.3
The force exerted by each muscle in Figure 9.2 to generate ankle plantarflexion moments, assuming a muscle coordination strategy that minimizes the sum of muscle forces. This objective function (Equation 9.1) results in the unrealistic behavior in which each muscle reaches its maximum force before the next muscle is recruited.

Although the objective function in Equation 9.1 leads to an optimization problem that can be easily solved, notice that only one muscle is recruited when a plantarflexion moment of 100 N·m is desired. If a greater moment is required, then we recruit a second muscle only when the first muscle has reached its maximum force-generating capacity. We know from experiments that muscles are not in fact recruited in this stepwise, noncooperative manner.

Different objective functions will lead to different solutions. More realistic objective functions have been proposed in the biomechanics literature but result in optimization problems that cannot be solved by inspection. Typically, we must use computers to solve the nonlinear, constrained, high-dimensional problems encountered in biomechanics. In the next sections, we will discuss two broad categories of optimization algorithms: *local methods* and *global methods*. We then move on to solve a more realistic muscle force-sharing problem.

Local methods to solve static optimization problems

Starting from an initial guess (candidate solution), a local optimization method generates a sequence of new guesses, each in the neighborhood of its predecessor and typically having a better objective function value. The algorithm terminates when it reaches a local optimum: a solution that is surrounded by inferior solutions in its immediate neighborhood. A classic example of a local method is the steepest-descent algorithm. If we are solving a minimization problem in two variables, for example, the objective function $J(x,y)$ can be represented as a surface whose height above the X–Y plane at the point (x^*, y^*) is $J(x^*, y^*)$. The steepest-descent strategy can then

FIGURE 9.4
Graphical representation of an objective function in two variables, where the height of the surface represents the objective function value $J(x,y)$. The steepest-descent algorithm starts at a point in the solution space and takes steps downhill, in the direction of the steepest local slope, until movement in any direction would increase $J(x,y)$. The final solution depends on the initial guess, as illustrated by the four paths shown here, and may not be globally optimal.

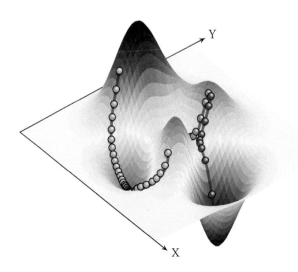
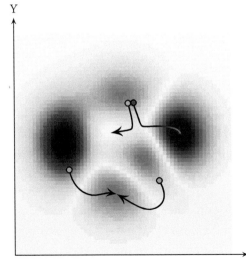

be imagined as placing a marble on the objective function surface and letting it roll downhill until it settles at the bottom of a basin. As shown in Figure 9.4, the final solution depends on the initial guess and may not be the best possible solution. Newton's method is similar to the steepest-descent method, but takes a more direct route to a local minimum by using information about the second derivative (curvature) of the objective function surface—though at the potentially substantial cost of computing this additional information. Interior-point methods are also popular, which begin by finding any feasible solution (i.e., a solution in the *interior* of the feasible region) and then seeking progressively better solutions while remaining within the feasible region.

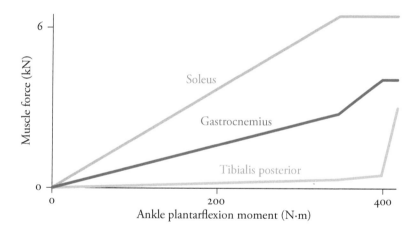

FIGURE 9.5
The force exerted by each muscle in Figure 9.2 to generate all possible ankle plantarflexion moments, assuming a muscle coordination strategy that minimizes the sum of squared muscle activations. This objective function (Equation 9.2) results in behavior that agrees with experimental observations, in which several muscles are recruited to generate even modest joint moments.

In Figure 9.5, we show the solution to Optimization Problem 1 (above) for all possible ankle plantarflexion moments when minimizing the sum of squared muscle activations, which has been used as a surrogate for metabolic power consumption:

$$J(\underline{F}) \triangleq \sum_{i=1}^{3} \left(\frac{F_i}{F_i^{\max}} \right)^2 = \sum_{i=1}^{3} a_i^2 \qquad (9.2)$$

Muscle activity as computed using Equation 9.2 is closer to what is observed in human experiments than when using the simpler objective function shown in Equation 9.1. In particular, notice in Figure 9.5 that all three muscles are recruited even when only small

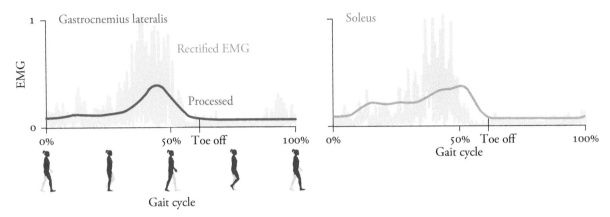

FIGURE 9.6
Electromyographic (EMG) signals from the gastrocnemius lateralis (left) and soleus (right) during walking. Both muscles contribute to generating ankle plantarflexion moments. Data from Bartlett and Kram (2008); Arnold et al. (2013).

ankle plantarflexion moments are desired, which is qualitatively similar to experimental measurements of EMG (Figure 9.6).

A local method seeks a solution that is optimal within its immediate neighborhood; however, there can be many such local minima. As illustrated in Figure 9.4, for example, the solution found by a gradient-descent algorithm depends on the initial guess. We might simply try running the algorithm a few times with different initial guesses, but this might still miss a solution that is better—perhaps substantially better. Remember Dick Fosbury. Thus, although local methods can be fast, a global method may lead to better solutions.

Global methods to solve static optimization problems

Global optimization methods avoid getting "stuck" in local minima by considering candidate solutions beyond the immediate neighborhood of the previous guess. One such algorithm is the genetic algorithm, where a population of candidate solutions evolves from one stage (generation) to the next through simulated natural processes. The philosophy behind this optimization strategy is to explore in more detail those regions of the solution space that are known to be favorable while also venturing into unexplored regions of the solution space that may contain even better solutions.

Genetic algorithms tend to scale poorly with problem size, but other evolutionary algorithms perform well in practice.

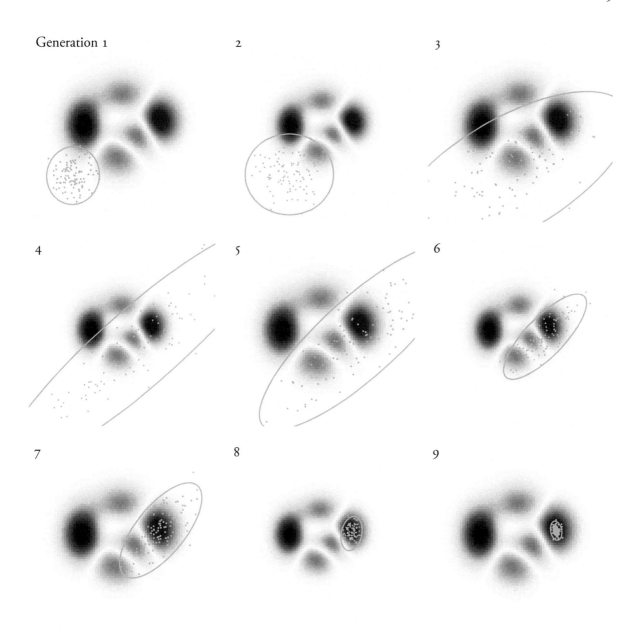

FIGURE 9.7
Global optimization methods like the covariance matrix adaptation evolution strategy (CMA-ES) avoid converging to local minima. Shown here are nine generations of the CMA-ES algorithm as it minimizes the function shown in Figure 9.4. Each cyan dot represents a candidate solution; the ellipse in each panel indicates the distribution of the population in that generation.

For example, the covariance matrix adaptation evolution strategy (CMA-ES) illustrated in Figure 9.7 performs very well for high-dimensional problems with many local minima. In each generation, the CMA-ES algorithm selects candidate solutions from a distribution whose mean and covariance are updated based on the favorability of the previous generations. Over time, the distribution will home in on a solution: the mean will approach an optimum and the distribution will shrink. Simulated annealing uses a similar idea; here the algorithm begins with a high "temperature" that permits more adventurous exploration of the solution space, then gradually "cools" into a more conservative steepest-descent method.

The best optimizer to use depends on the nature of the problem. For example, if the objective function is smooth and the problem has only one minimum, then a simple local search strategy will suffice as it will be guaranteed to converge to the global minimum. On the other hand, algorithms like CMA-ES would be more appropriate in the presence of many local minima. For large problems, it is advantageous to use an algorithm that can be run in parallel and to define criteria for deciding when a solution will be good enough to answer a particular research question.

Muscle forces during walking and running

With these optimization methods in hand, we are now equipped to solve one of the most important problems in biomechanics: determining the muscle forces that are responsible for generating movements such as walking and running. This is called the muscle redundancy (or force-sharing) problem.

When we run, our muscles generate hip extension, knee extension, and ankle plantarflexion moments during the first part of the stance phase. Suppose we wish to compute the muscle forces generated at the instant of peak vertical ground reaction force (Figure 9.8). Using the lower-limb model and data given in

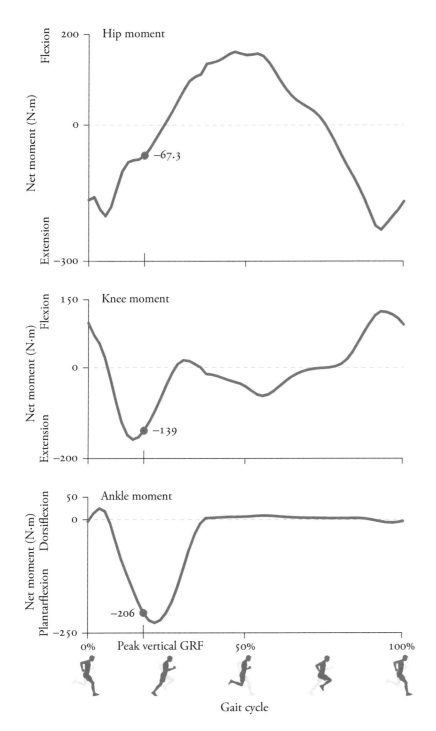

FIGURE 9.8
Net joint moments generated in the sagittal plane by one subject while running at 5 m/s. The vertical line indicates the time of peak vertical ground reaction force (GRF). Data from Hamner and Delp (2013).

Figure 9.9, and the objective function given in Equation 9.2 to minimize the sum of squared muscle activations, we can express the optimization problem as follows:

> **Optimization Problem 2:** Find muscle activations at the instant of peak vertical ground reaction force during running
>
> minimize $\quad J(\underline{a}) = \sum_{i=1}^{9} a_i^2$
>
> subject to $\quad \sum_{i=1}^{9} a_i \left(r_i^{hip} F_i^{max} \right) = -67.3$
>
> $\quad\quad\quad\quad\ \sum_{i=1}^{9} a_i \left(r_i^{knee} F_i^{max} \right) = -139$
>
> $\quad\quad\quad\quad\ \sum_{i=1}^{9} a_i \left(r_i^{ankle} F_i^{max} \right) = -206$
>
> $\quad\quad\quad\quad\ 0 \leq a_i \leq 1 \quad \text{for } i = 1, \ldots, 9$

The solution to this optimization problem estimates the forces generated by each muscle in the model at one point in time:

Muscle or group	Force, F_i (N)
Gluteus maximus	875
Iliopsoas	0
Hamstrings	340
Rectus femoris	0
Biceps femoris short head	0
Vasti	4134
Gastrocnemius	1396
Soleus	4167
Tibialis anterior	0

MUSCLE FORCE OPTIMIZATION   233

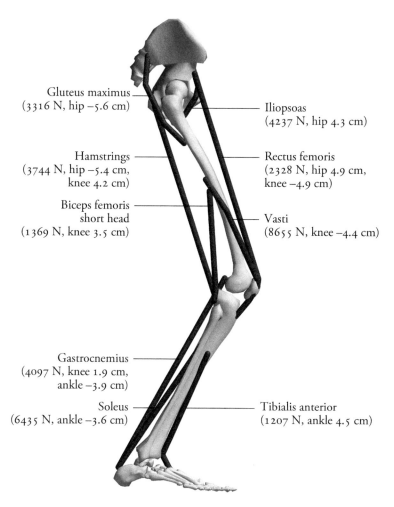

FIGURE 9.9
A simple musculoskeletal model of the leg can be used to study muscle coordination and joint loads during the stance phase of running. Key muscles involved in generating sagittal-plane movement have been grouped into nine representative muscle paths (Hamner and Delp, 2013). Values in parentheses are the instantaneous maximum force (assuming zero velocity and a rigid tendon) and moment arms corresponding to the pose shown.

We can estimate the forces generated by each muscle during walking and running by repeating this analysis at evenly spaced instants over the gait cycle. The muscle forces and resulting joint moments are shown for walking in Figures 9.10 and 9.11; for running, in Figures 9.12 and 9.13. As expected, the computed muscle forces are higher during running than during walking. Also notice that the muscle forces peak earlier in the gait cycle during running, as we would expect given the shorter stance phase.

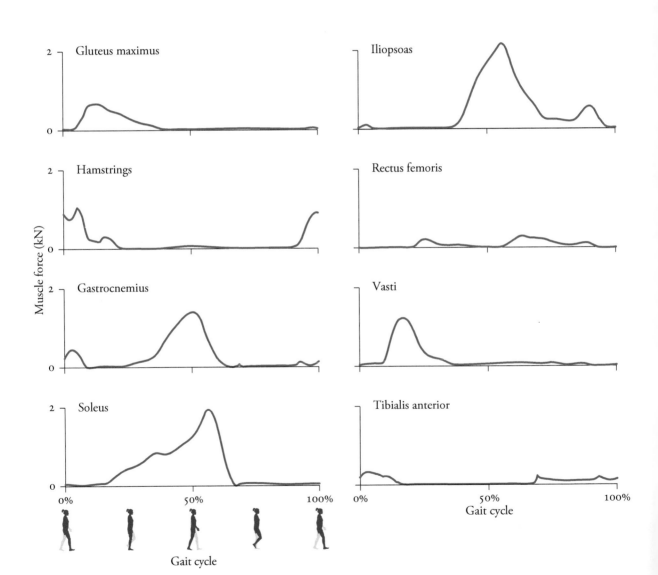

FIGURE 9.10
Muscle forces for one subject (male, 67.1 kg) walking at freely selected speed (1.67 m/s), computed using Static Optimization in OpenSim. Data from Dembia et al. (2017).

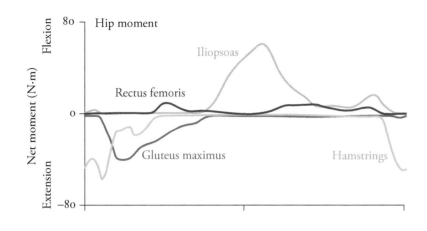
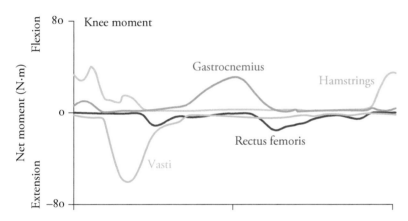
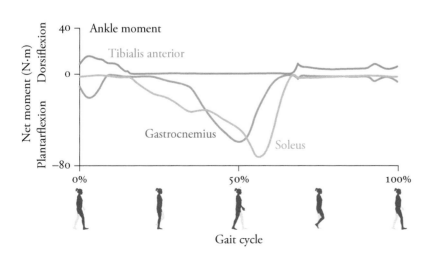

FIGURE 9.11
Sagittal-plane joint moments generated during walking at 1.67 m/s by the muscle forces shown in Figure 9.10.

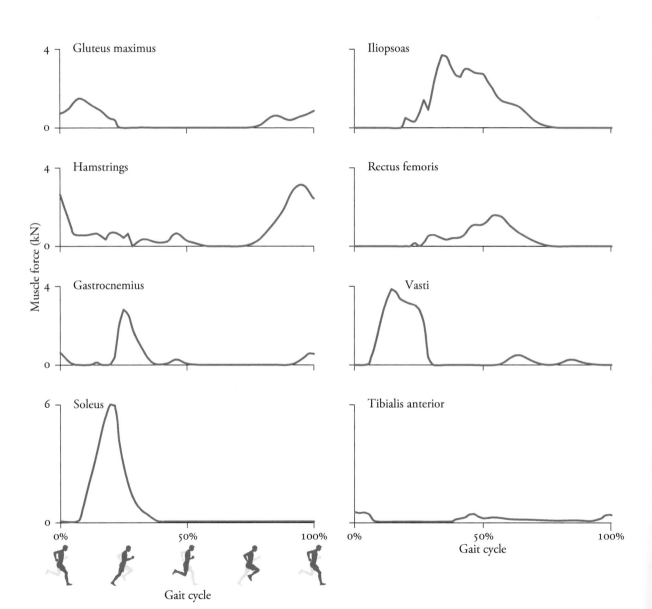

FIGURE 9.12
Muscle forces for one subject (male, 69.4 kg) running at 5 m/s, computed using Static Optimization in OpenSim. Data from Hamner and Delp (2013).

MUSCLE FORCE OPTIMIZATION   237

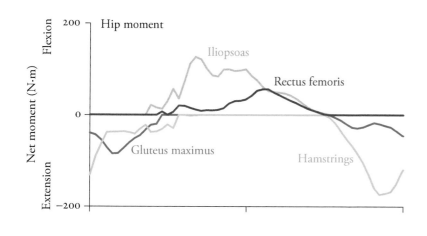
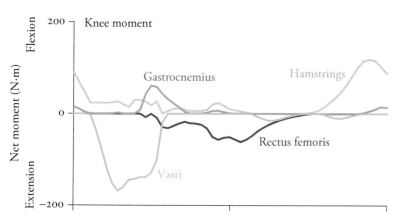
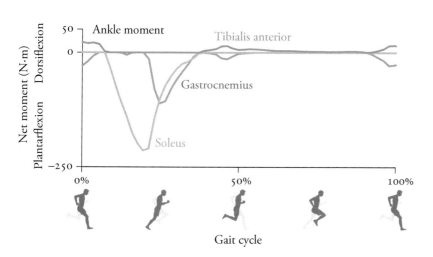

FIGURE 9.13
Sagittal-plane joint moments generated during running at 5 m/s by the muscle forces shown in Figure 9.12.

Estimating joint loads

Joint loads can only be estimated once muscle forces are known. As we mentioned in Chapter 8, it is important to distinguish between the net joint reaction forces computed from an inverse dynamic analysis and the "bone-on-bone" forces that would be measured by a sensor implanted in a joint. The intersegmental forces we computed in Chapter 8 would be equal to the actual joint loads only if our muscles applied pure joint torques like rotational motors. Of course, our muscles do not generate torques directly but rather generate pulling forces that are applied to the skeleton. These forces produce moments as well as compressive and shearing forces in the joints. As we will see, the contribution of muscle forces to joint loads can be substantial.

To demonstrate how muscle forces contribute to joint loads, we consider the instant at which the vertical ground reaction force peaks during running at 5 m/s. For simplicity, we will use the planar model shown in Figure 9.14 to compute the compressive and shearing forces at the ankle. In the previous section, we used the lower-limb model shown in Figure 9.9 to estimate the forces that would be generated by the gastrocnemius and soleus at this instant. Note, however, that the paths of the gastrocnemius and soleus muscles in that model do not lie in the plane perpendicular to the ankle joint axis, so we retain in our forthcoming calculations only the components of these forces that act in the plane of Figure 9.14. We project the ground reaction force onto this plane as well. Finally, we assume the foot is stationary during the stance phase and compute forces Fx and Fy as in Chapter 8:

$$Fx = F_x^{GAS} + F_x^{SOL} + F_x^{GRF}$$
$$= 44\text{ N} + 495\text{ N} - 838\text{ N} \qquad (9.3)$$
$$= -299\text{ N}$$

$$Fy = F_y^{GAS} + F_y^{SOL} + F_y^{GRF}$$
$$= 1357\text{ N} + 4088\text{ N} + 1379\text{ N} \qquad (9.4)$$
$$= 6824\text{ N}$$

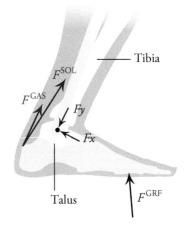

FIGURE 9.14
Planar model used to estimate ankle joint loads during running at 5 m/s, at the instant of peak vertical ground reaction force (GRF). The forces shown have been projected into the plane perpendicular to the ankle joint axis.

Notice that the muscle forces (F^{GAS} and F^{SOL}) contribute substantially to the compressive force in the joint (Fy), accounting for about 80 percent of the total. The total compressive force is over 10 bodyweights in this example during running, but substantial compressive loads have been measured in less dynamic activities as well. Instrumented artificial joints have recorded peak contact forces in the hip of about 2.5 bodyweights during walking and over 5 bodyweights during periods of instability while standing on one leg. Even during a stationary single-leg stance, the contact force in the hip is over 2 bodyweights due to the forces generated by muscles. Thus, it is essential to calculate muscle forces if we wish to assess the forces within joints.

Dynamic optimization

In the preceding sections, we made several assumptions about the underlying optimization problem being solved. Perhaps most importantly, we assumed that the objective function depends only on instantaneous quantities. We ignored muscle–tendon dynamics and assumed that muscle activations (and the consequent forces that are generated) depend only on the net joint moments required at a particular instant. Static optimizations such as these are relatively easy to handle computationally, but we know that humans and animals anticipate the future. For example, we may invest energy at the beginning of a movement if it will be profitable later, as when we "wind up" before throwing a ball or squat before performing a standing long jump. These explosive activities take advantage of the dynamics of our muscles and tendons. Static optimization may be insufficient to fully understand muscle coordination during movements such as throwing, jumping, and sprinting, where accurately predicting the muscle activations at one point in time may require us to consider the entire movement.

A second key assumption we made in the previous sections is that the kinematics and net joint moments are known *a priori*. If we are studying a movement for which motion capture data are available, for example, we can estimate joint motions and moments computationally using inverse kinematic and inverse dynamic simulations. However, suitable experimental data may be unavailable and difficult (or impossible) to collect, such as when studying activities with high injury rates or the gait of *Tyrannosaurus rex*. Further, we may specifically wish to predict unobserved movement adaptations in response to surgeries, implants, prostheses, or exoskeletons, or to discover novel movement strategies for maximizing athletic performance or minimizing joint loads. We can use dynamic optimization for these tasks.

In a dynamic optimization, we use a model of musculoskeletal dynamics to determine the muscle coordination and consequent motion that optimize a mathematical description of a motor task. Dynamic optimization typically proceeds by selecting a candidate solution, running a forward dynamic simulation, evaluating the performance of the candidate solution, and iterating until some stopping criterion is met (e.g., a target objective function value has been reached or a predetermined number of candidate solutions has been considered; Figure 9.15). The design variables might include musculoskeletal parameters like the length of the Achilles tendon, parameters of a wearable device like the stiffness of a performance-enhancing shoe, or the activity of each muscle over time.

For example, suppose we wish to optimize muscle coordination during a 100 m sprint. Each candidate solution might describe the activity of each muscle over time (e.g., as a parameterized curve or sequence of control points). We would seek to minimize the time required to travel 100 m, perhaps including constraints to prevent injury. Evaluating the objective function would involve running a forward dynamic simulation (see Figure 1.12) where the model is posed in an initial configuration, the proposed activations at the first time point are applied to the model's muscles, the resulting muscle forces are applied to the skeleton, the consequent accelerations ($\underline{F} = m\underline{a}$) are numerically integrated forward in time to determine the model's configuration at the next time point, and the process is repeated until the runner reaches the finish line. Solving this optimization problem would require substantial computational resources. To make the problem more tractable, we might use a

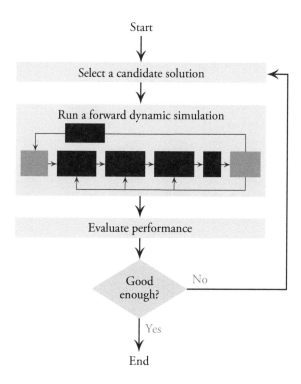

FIGURE 9.15
A dynamic optimization proceeds by selecting a candidate solution, running a forward dynamic simulation, evaluating the performance of the candidate solution, and iterating until the stopping criterion is met. Candidate solutions can be selected by any optimizer, may adjust any parameters in the model, and may be evaluated based on any performance metrics.

planar model that includes only a few key muscles, focus on a single gait cycle at a fixed speed, or design a neural controller and optimize a relatively small number of controller parameters rather than the activation of each muscle at each point in time.

The objective function and constraints can take many forms, depending on the research question being explored. In the sprinting example above, we simply sought to minimize duration. For a task like walking to the bus stop or opening a door, however, it would be more appropriate to use an objective function that combines minimizing duration with minimizing total energy expenditure, perhaps balancing these terms with weights that reflect the urgency of the activity. Analysis of other tasks might demand additional terms that account for considerations like safety, comfort, fatigue, and stability. Note that optimizers are "clever" and will exploit any aspect of an optimization problem that leads to a better objective function value, regardless of the subjective nature of the solution—so be careful what you wish for! I once defined an objective function to minimize the cost of transport without imposing a minimum travel distance. The optimizer wisely chose to simply fall forward, which took very little energy but was not what I intended. Defining optimization problems can be an iterative process, where the solutions returned by the optimizer are used to identify errors in the translation from research question to objective function and constraints.

As is the case with any research study, it is critical to validate the results obtained through optimization. As we saw in Figure 9.6, the timing of computed muscle activity might be compared to the timing of experimental EMG. Experimental kinematics, ground reaction forces, joint loads (e.g., from instrumented joint replacements), indirect calorimetry, and other measurements for the same or similar activities can also be useful. Robust, properly validated results from dynamic optimizations can lead to deep insights about movement and muscle coordination. For example, B. J. Fregly used optimization to discover a new way to walk that reduces the loads in his knees (Fregly et al., 2007). He learned how to walk with this gait and, as a result, was able to keep up with his kids when they visited Disney World.

In the case of the Fosbury flop, a dynamic optimization might reveal that the jumper should raise his arms as he leaves the ground to maximize the vertical velocity of his mass center at takeoff. Likewise, he should maintain contact with the ground as long as possible during takeoff to maximize the impulse of the ground reaction force. These insights have been obtained empirically, through the experience of athletes and coaches over the 50 years since Fosbury debuted his flop. In the next section, by contrast, we will describe an example in which such insights have been achieved much more rapidly through computer modeling and numerical optimization.

Muscle coordination during a standing long jump

Assistive devices such as orthoses, prostheses, and exoskeletons have the potential to improve human life by restoring mobility following injury and enhancing athletic performance. However, we do not yet fully understand how assistive devices interact with the coordination of our muscles when we are performing complex tasks. My student Carmichael Ong set out to investigate how a simulated assistive device could improve performance of a standing long jump. This task demands precise muscle coordination and has a clear objective, and is therefore well suited for optimization.

Carmichael used a planar, five-segment model (Figure 9.16). Physiological torque actuators at the ankle, knee, hip, and shoulder represented the combined action of all muscles crossing each joint and included the length- and velocity-dependent force-generation capacities found in biological muscle. The activity of each torque actuator was described by a piecewise linear function, the nodes of which were the design variables in the dynamic optimization problem. The objective function in this problem maximized jump distance (by minimizing its negative) while avoiding undesirable outcomes:

$$\begin{aligned}\text{minimize} \quad J = &-d & &\text{Reward longer distances} \\ &+ w_1(J_{\text{fall}}) & &\text{Penalize falls at landing} \\ &+ w_2(J_{\text{injury}}) & &\text{Discourage use of ligaments} \\ &+ w_3(J_{\text{slip}}) & &\text{Penalize slipping at takeoff} \\ &+ w_4(J_{\text{time}}) & &\text{Reward counter-movements}\end{aligned} \quad (9.5)$$

Weights w_i in Equation 9.5 were chosen to reflect the relative importance of each term and to appropriately balance units of length,

FIGURE 9.16
A planar model with five segments and seven degrees of freedom for studying muscle coordination during a standing long jump. Dynamic optimization was used to predict the coordination of muscles crossing the ankle, knee, hip, and shoulder to maximize jump distance. Adapted from Ong et al. (2016).

torque, force, and time. Determining appropriate weights in the objective function is often an iterative process, with adjustments being made manually or using another optimizer (in a process sometimes called "meta-optimization").

In this study, Carmichael began by using the CMA-ES algorithm to optimize the jump distance of the model without assistance. This initial optimization problem produced a standing long jump that captured the salient kinematic and kinetic features of human jumps, including a counter-movement, takeoff kinematics, ground reaction forces, and joint moments. He then augmented the model with massless rotational springs placed at the hip, knee, and ankle (Figure 9.17). Of course, such devices do not exist, but the simulations allowed us to adjust the applied torques in any way we wished, which helped us to understand how they affect performance.

In the augmented scenario, Carmichael optimized the stiffness and equilibrium position of each spring simultaneously with the other design variables. The optimal assistive device increased jump distance by over 1 m, from 2.27 m to 3.32 m. It was exciting to see the algorithm co-optimize the device parameters and the human coordination so that their combined performance achieved the longest possible jump. This study demonstrates how dynamic optimization can provide insight into human movement and complement experimental approaches to assistive device design. In the future, predictive tools like this may be used to customize a surgery, implant, prosthesis, or assistive device for an individual patient, perhaps helping to determine which of these treatments would lead to the best outcome.

FIGURE 9.17
Joint torques during the ground contact phase when unassisted (left) and augmented (right) that maximize performance of a standing long jump. Mean ±1 standard deviation shown for experimental data (Horita et al., 1991). Adapted from Ong et al. (2016).

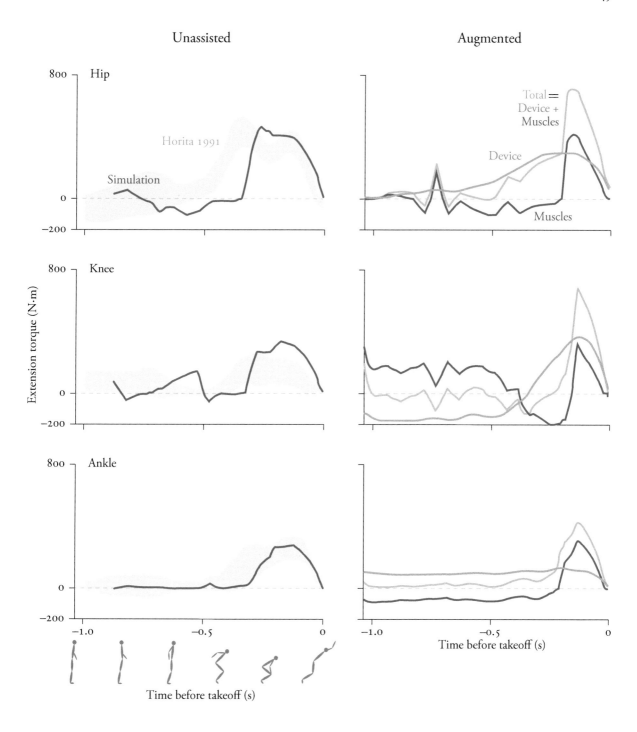

Part IV
Muscle-Driven Locomotion

10 Muscle-Driven Simulation

Prediction is very difficult,
especially if it's about the future.
—Niels Bohr

MY FIRST JOB AFTER GRADUATING from college in 1983 was helping small companies write computer-aided design software. I was working at a computer factory in Colorado that had just produced a powerful new graphics computer. Powerful in those days meant that it could draw a few lines on a small screen every second. No one would use our graphics computers without graphics software, so my job was to help engineers at other companies get their computer-aided design software running on our newly minted computers. What became clear to me was that nearly all future products would be designed on computers. In my applications for graduate school, I proposed building computer graphics tools for designing surgeries.

When I entered graduate school two years later, I was fortunate to join a biomechanics research laboratory led by Felix Zajac in the Design Group at Stanford University. As a neuroscientist interested in understanding the control of movement, Zajac had spent decades performing elegant experiments to measure muscle activation patterns, ground reaction forces, and joint motions during movements such as jumping. He realized that synthesizing these experimental data into a comprehensive understanding of muscle function during movement would require more than expert analysis of the data.

Experimental measurements alone are insufficient to understand muscle actions during movement for two reasons. First, important quantities like the forces generated by muscles cannot usually be measured in experiments. Second, it is difficult to establish cause–effect relationships through experimental observation alone. For example, ground reaction forces can be measured during walking (Chapter 2) and can be used to estimate the accelerations of the body's center of mass. However, ground reaction force measurements alone offer little insight into how muscles contribute to the accelerations of the body's center of mass, and thereby to the critical tasks of supporting the body's weight and propelling the body forward during walking. EMG signals (Chapter 4) can be analyzed to understand when muscles are active but do not reveal which motions of the body arise from each muscle's activity.

We needed a new framework to advance our understanding of muscle function during movement. This framework had to reveal the relationships between muscle activations, muscle forces, ground reaction forces, and motions of the body.

Muscle-driven simulations of movement offer this framework. Simulations provide estimates of muscle forces and reveal cause–effect relationships, such as a muscle's contribution to the ground reaction force during walking. We can also use simulations to predict how the body will respond to a disease, surgery, or altered muscle activations. These capabilities allow us to characterize the actions of muscles during movement and to design surgeries and assistive devices.

Zajac and his students were at the forefront of developing muscle-driven simulations when I entered his research group in 1985. Joining this group started me on a journey lasting over 30 years that has focused on creating muscle-driven simulations and analyzing them to improve movement for people with impaired biomechanics. Whereas my work just after college produced computer-aided design tools for engineered products within two years, developing computer-aided design tools for understating the complexities of human movement turned out to be a lifelong challenge.

This chapter describes some of what I have learned about creating muscle-driven simulations. The chapter begins by illustrating why it is so difficult to determine the actions of muscles during movement without a simulation, and why the literature is riddled with incorrect conclusions about muscle function. We move on to discuss the four stages of building and analyzing muscle-driven simulations to determine muscle actions correctly. The chapter then describes the open-source simulation software my colleagues and I have developed to enable a worldwide collaboration in which thousands of researchers build and share computer simulations of movement. My goal is to advance the field by pulling together.

Understanding muscle actions during movement is challenging

Experimental approaches to infer a muscle's actions based on its geometry, EMG measurements, and observed motions do not correctly explain how muscles move the body. Analyses based on anatomical knowledge alone frequently lead to incorrect conclusions about the function of muscles. For example, the soleus muscle is described in much of the anatomical and biomechanical literature as a muscle that plantarflexes the ankle. The soleus muscle does generate an ankle plantarflexion moment which does, indeed, plantarflex the ankle, but the muscle can also perform other actions (Figure 10.1). These actions arise from an effect known as *dynamic coupling*.

Dynamic coupling describes the phenomenon whereby the motion of one body segment affects the motion of another segment due to induced forces. As shown in Figure 10.1, the force generated by the soleus not only generates an ankle plantarflexion moment but also induces intersegmental forces and joint accelerations throughout the body. The magnitudes and directions of these intersegmental forces depend on the force applied by the muscle, the muscle's moment arm, the mass and inertia of the body segments, and the pose of the body. In the example shown on the right side of Figure 10.1, the force generated by the soleus produces a counterclockwise angular acceleration of the shank, which requires the knee joint to accelerate upward and to the left. The inertia of the thigh and adjoining segments resists this acceleration and results in an intersegmental force at the knee—which in turn accelerates the

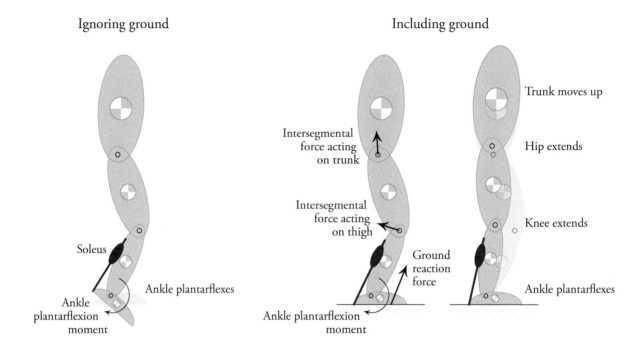

thigh, and so on. Thus, the soleus accelerates all the joints of the body despite spanning only the ankle.

In many cases, the intersegmental forces resulting from dynamic coupling are large enough to influence our interpretation of a muscle's actions. Although the "muscle-induced" accelerations are generally small at joints far from the muscle, they can be substantial at nearby joints. For example, Felix Zajac and Michael Gordon (1989) demonstrated that, during standing, the soleus can accelerate the knee into extension even more than it accelerates the ankle into plantarflexion. Furthermore, they noted that it is possible for a biarticular muscle to induce a joint acceleration that opposes the moment it generates about one of the joints it crosses. For example, although the gastrocnemius generates a knee flexion moment and an ankle plantarflexion moment, it can nevertheless induce a knee extension acceleration or an ankle dorsiflexion acceleration (Figure 10.2). These seemingly incongruous accelerations are possible when the gastrocnemius is activated because, for example, the knee flexion acceleration due to the knee flexion moment it generates may be

FIGURE 10.1
Actions of the soleus muscle during single-limb stance, ignoring (left) and including (right) dynamic effects. The ankle plantarflexion moment generated by the soleus plantarflexes the ankle in both cases but is seen to affect other joints and body segments when the ground reaction force is considered, due to dynamic coupling. Adapted from Anderson et al. (2006).

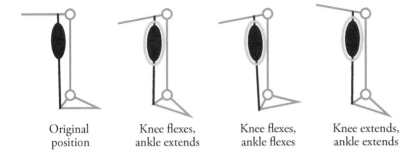

FIGURE 10.2
A biarticular muscle can induce a joint acceleration that opposes the moment it generates about one of the joints it crosses. For example, from the initial position shown (far left), a force generated by the gastrocnemius may act to flex or extend the knee and ankle as shown due to dynamic coupling. Adapted from Zajac (1993).

overshadowed by the knee extension acceleration induced by the ankle plantarflexion moment it generates. Many biomechanical studies have ignored dynamic coupling when interpreting muscle actions and have drawn erroneous conclusions. It is challenging to deduce the actions of muscles during movement for musculoskeletal systems that consist of dozens of body segments, joints, and muscles. Muscle-driven simulations are needed to meet this challenge.

You may be wondering whether muscles really produce accelerations in the direction opposite to their applied moments. I had my doubts as well. Steve Piazza was a student working in my lab who had created muscle-driven simulations of the swing phase of walking (Piazza and Delp, 1996). His simulations showed that the hamstrings generated a hip flexion acceleration in some situations. Recall that the hamstrings cross behind the hip and, for this reason, many scientists believe that these muscles always produce hip extension. Analysis of the equations of motion confirmed that generation of hip flexion was indeed possible, but our clinical colleagues were skeptical. In particular, Jacqueline Perry, a world-leading expert in muscle and gait, and one of my scientific heroes, did not believe our results and wanted more evidence. So Steve built "the convincer," a simple mechanism that resembled a leg with a wire where the hamstrings would be. When we pulled on the hamstrings wire under the right conditions, the hip would flex slightly. Steve and I were convinced—and so were our clinical colleagues, including Dr. Perry.

Creating muscle-driven simulations

The equations arising from Newton's laws of motion characterize the dynamics of our bodies. We can make predictions about how a person will move by solving these equations over time, a process referred to as *dynamic simulation*. A "muscle-driven" dynamic simulation predicts how the forces produced by muscles contribute to motions of the body segments during movements such as walking and running.

The process of developing, testing, and analyzing muscle-driven simulations comprises four stages (Figure 10.3). In Stage 1, you create a computational model that characterizes the dynamic behavior of the musculoskeletal system with sufficient accuracy to answer your research question. You may be able to skip this laborious step if someone else has created and shared a model that is appropriate for your research. Stage 2 involves computing a set of muscle excitations which, when applied to the model, generate a simulation of the movement of interest. Stage 3 confirms that the simulation adequately represents the movement of interest by comparing the simulation results to experimental measurements. In Stage 4, you analyze your simulations to answer your research question. We explore each of these stages in the sections that follow.

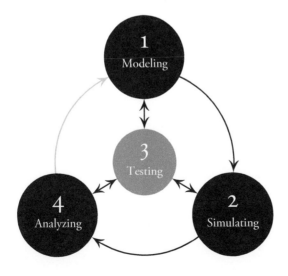

FIGURE 10.3
The process for creating and analyzing muscle-driven simulations includes (1) modeling musculoskeletal dynamics, (2) simulating movement, (3) testing the accuracy of the simulations, and (4) analyzing the simulations to answer a specific research question.

Stage 1: Modeling musculoskeletal system dynamics

A model of musculoskeletal dynamics allows us to compute the motions caused by each muscle force. The first stage of our four-stage process is to create a model of the musculoskeletal system using equations that describe muscle activation dynamics, muscle–tendon contraction dynamics, musculoskeletal geometry, and skeletal dynamics (Figure 10.4). These equations characterize the time-dependent behavior of the musculoskeletal system in response to muscle excitations.

As we saw in Chapter 4, the relationship between muscle excitation and activation is governed by the dynamics of the motor unit action potentials and cross-bridge cycling. The activation of a muscle (a) can be modeled by relating its time derivative ($\dot{a}$) to the current activation and excitation (u), as described by Equation 4.1.

Muscle activations are inputs to the model of muscle–tendon contraction dynamics, as are the lengths and lengthening velocities

FIGURE 10.4
Elements of a muscle-driven simulation. Excitations from a neural controller produce muscle activations and muscle forces via models of muscle activation and contraction dynamics. These forces are transmitted to the skeleton, generating joint moments that accelerate the body segments. Sensory feedback modifies neural commands. Feedback is also shown to indicate that body movement affects other elements of the system.

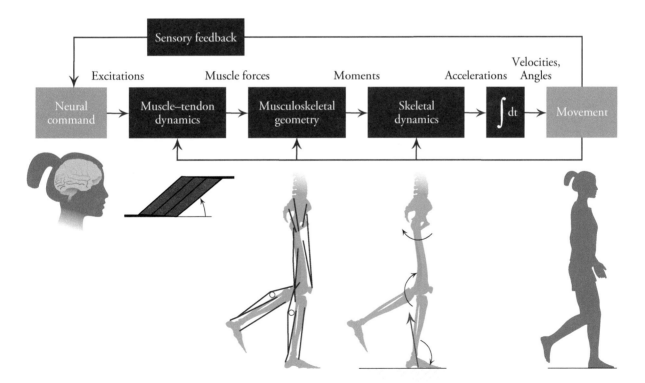

of the muscle–tendon actuators ($\underline{\ell}^{MT}$ and $\underline{v}^{MT}$). As described in Chapter 5, the dynamics of muscle and tendon are governed by the time course of cross-bridge formation, the sliding of actin filaments, and the dynamics of tendon. The forces generated by a muscle (F^M) and transmitted through its tendon (F^T) can be estimated using four dimensionless curves and five muscle-specific parameters, as detailed in Chapter 5. When applied to the skeleton, muscle forces produce moments about the joints, as discussed in Chapter 6. The joint moments generated by muscles cause the joints and body segments to accelerate, thereby producing movement.

The accelerations of the body in response to muscle forces and other loads can be computed using the equations of motion for the body:

$$\underline{\ddot{q}} = M^{-1}(\underline{q})\left\{\underline{F}^G(\underline{q}) + \underline{F}^C(\underline{q},\underline{\dot{q}}) + R(\underline{q})\,\underline{F}^T(\underline{u}) + \underline{F}^E(\underline{q},\underline{\dot{q}})\right\} \quad (10.1)$$

In this set of equations, $\underline{q}$, $\underline{\dot{q}}$, and $\underline{\ddot{q}}$ are the positions, velocities, and accelerations of the joints. $M^{-1}(\underline{q})$ is the inverse of the system mass matrix, which contains the masses and inertial properties of the body segments. As shown in Equation 10.1, the mass matrix defines the relationships between forces and joint accelerations. $\underline{F}^G(\underline{q})$ is a vector of forces arising from gravity. $\underline{F}^C(\underline{q},\underline{\dot{q}})$ is a vector of Coriolis and centrifugal forces, which arise when Newton's laws of motion are applied in reference frames that are fixed to rotating bodies. $R(\underline{q})$ is a matrix of muscle moment arms and $\underline{F}^T(\underline{u})$ is a vector of tendon forces, which is a function of muscle excitations ($\underline{u}$). The muscle moment arms determine the joint moments that result when the tendon forces are applied to the skeleton. Finally, $\underline{F}^E(\underline{q},\underline{\dot{q}})$ is a vector of external forces that characterize the interactions between the body and its environment (e.g., ground reaction forces). Analyzing Equation 10.1 allows us to compute the accelerations caused by each muscle force and thereby determine the actions of a muscle in a complex multijoint system.

The modeling approach illustrated in Figure 10.4 has been used to create muscle-driven simulations of a wide variety of movements. Some studies use a model of the musculoskeletal system that is relatively simple. For example, a two-dimensional

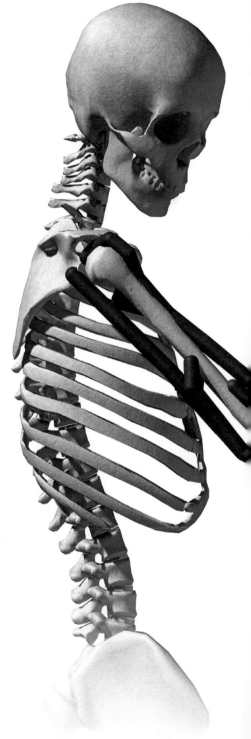

model with only a few degrees of freedom and actuated by just a few muscles may be sufficient to capture the phenomena of interest (Figure 10.5). In other cases, a three-dimensional model with many degrees of freedom and dozens of muscles may be necessary to elucidate the contributions of individual muscles to an observed movement (Figure 10.6). Note that the usefulness of a model does not always increase as model complexity increases. In fact, it is counterproductive to use a model that is more complex than is necessary to answer your research question. Additional complexity increases the effort required to build and validate a model, generate simulations, and analyze simulation results. So beware: complexity begets perplexity!

Once a generic model has been selected for a particular study, the model must be calibrated to match the geometry and strength of each study participant. You can calibrate a model by adjusting

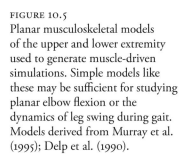

FIGURE 10.5
Planar musculoskeletal models of the upper and lower extremity used to generate muscle-driven simulations. Simple models like these may be sufficient for studying planar elbow flexion or the dynamics of leg swing during gait. Models derived from Murray et al. (1995); Delp et al. (1990).

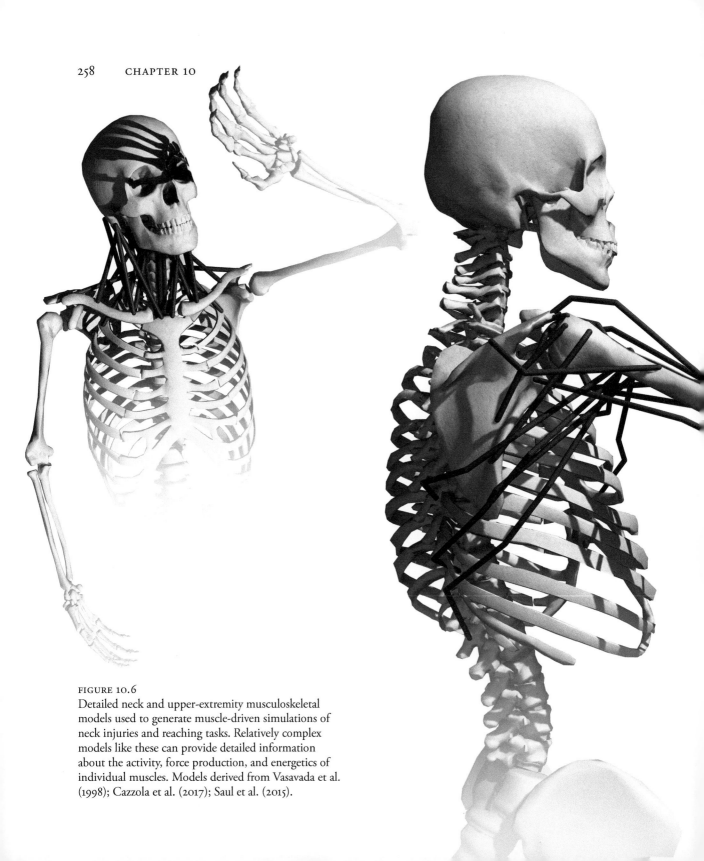

FIGURE 10.6
Detailed neck and upper-extremity musculoskeletal models used to generate muscle-driven simulations of neck injuries and reaching tasks. Relatively complex models like these can provide detailed information about the activity, force production, and energetics of individual muscles. Models derived from Vasavada et al. (1998); Cazzola et al. (2017); Saul et al. (2015).

the size and inertial properties of each body segment based on measurements made during a motion capture experiment (see Chapter 7). The location, orientation, and range of motion of the model's joints may be adjusted based on experiments in which the subject moves each joint through a series of standard motions. The paths and other parameters of muscles may be calibrated using clinical exams, strength tests, and activities like a maximum-height jump. A powerful approach to calibrate a model is to adjust model parameters through optimization, minimizing the error between a measurement and an analogous simulation.

Although we have spent only a few pages here, it can take years to build, calibrate, and validate a new musculoskeletal model. I urge model developers to share their creations with the biomechanics community so that others can build upon their work.

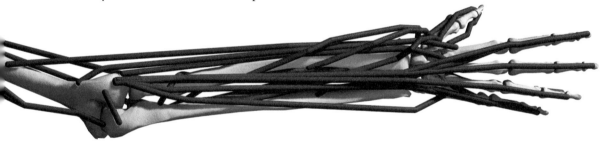

Stage 2: Simulating movement

To generate a simulation of movement, the differential equations of a dynamic model are numerically integrated forward in time. If the simulation is driven by muscles, a set of muscle excitations (i.e., $\underline{u}$ in Equation 10.1) must be applied. The muscle excitations are usually generated by an optimizer. One must also provide the *initial conditions* of the simulation, or the value of each state variable at the initial time. State variables include the joint angles and angular velocities, muscle activations, muscle fiber lengths, and other time-varying quantities. The governing differential equations (Equation 10.1) describe the rate at which these state variables change as a function of their current values. The numerical solution of these equations yields the trajectories of the state variables, from which one can compute all other state-dependent quantities, such as muscle and tendon forces, tendon strains, joint contact forces, and the metabolic energy consumed by each muscle.

Finding a set of muscle excitations that produces a coordinated movement is challenging, particularly for movements as complex as walking. Not only must many degrees of freedom be controlled, but the time-dependent, nonlinear force-generating properties of muscle must also be taken into account. Furthermore, we are often interested in generating simulations that agree with experimental measurements, such as joint angle trajectories or metabolic power consumption over time. To overcome these challenges, dynamic optimization (Chapter 9) can be used to find muscle excitations that minimize a specified performance criterion. Muscle excitations can be parameterized in a variety of ways. A common approach is to represent each excitation signal as a sequence of points spaced evenly over time, where the excitation at an instant in time is computed from the two neighboring points using linear interpolation (Figure 10.7). In this case, the optimizer would maximize performance by adjusting the height of each point.

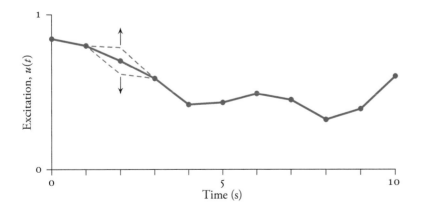

FIGURE 10.7
A curve representing the excitation of a muscle can be parameterized as a sequence of points over time. An optimizer predicts muscle coordination during movement by solving for the height of each point for each muscle.

Two optimization strategies can be used to generate muscle-driven simulations. One approach is to solve an optimal tracking problem, where the objective function reflects the difference between simulated and measured quantities such as joint angles, joint powers, and ground reaction forces. By minimizing the objective function over the duration of the simulation, the model is driven to reproduce a movement that was observed experimentally. For example, Anne Silverman and Richard Neptune (2012) used the following objective function to generate muscle-driven simulations of amputee gait that track experimental measurements:

$$J = \sum_i \sum_j \frac{(y_{ij} - \hat{y}_{ij})^2}{\sigma_i^2} \quad (10.2)$$

where y_{ij} is the measurement of variable i (including joint kinematics and the components of the ground reaction force vectors) at time step j, $\hat{y}_{ij}$ is the corresponding quantity from the simulation, and σ_i^2 is the variance of experimental quantity y_i. The squared errors are divided by the variance of the corresponding experimental variable so that variables with large variance are tracked less closely.

Solving optimal tracking problems can be computationally expensive. Even simple problems like tracking kinematics for a single gait cycle may take hours or days to solve. I was fortunate to work with Darryl Thelen and Clay Anderson when they were developing the computed muscle control algorithm, which solves the optimal tracking problem 100–1,000 times faster than previous approaches (Thelen et al., 2003). Many people now use this approach.

The second method to generate muscle-driven simulations of movement is dynamic optimization, where movements are generated using goals like minimizing total energy consumption or maximizing a metric of athletic performance. As we saw in Chapter 9, Carmichael Ong and colleagues generated simulations of a standing long jump using an objective function that rewards longer distances and penalizes undesirable solutions, such as those that would cause injuries. Tim Dorn, who was also working in my lab at the time, used an objective function of a similar form to generate muscle-driven simulations of loaded and inclined walking:

$$J = w_{fall} J_{fall} + w_{speed} J_{speed} + w_{head} J_{head} + w_{effort} J_{effort} \quad (10.3)$$

where J_{fall} penalizes unstable gaits, J_{speed} penalizes gaits that differed from a target speed, J_{head} penalizes gaits that involved unrealistic head motions, J_{effort} penalizes gaits that were energetically expensive, and weights w_i balance these competing objectives. Solving this dynamic optimization problem produces simulations of walking in a variety of scenarios, including simulations of humans walking uphill while carrying heavy loads (Dorn et al., 2015).

Stage 3: Testing the accuracy of dynamic simulations

Once you have developed a model of a musculoskeletal system, you must confirm that the behavior of the system is represented with sufficient accuracy to answer your research questions. This process is known as *validation*. My colleague Jennifer Hicks led our team in writing a review article that addresses the question faced by all of us involved in computer simulation: Is my model good enough? (Hicks et al., 2015).

All models are designed with intended uses and have limitations. It is important to be aware of these limitations when selecting

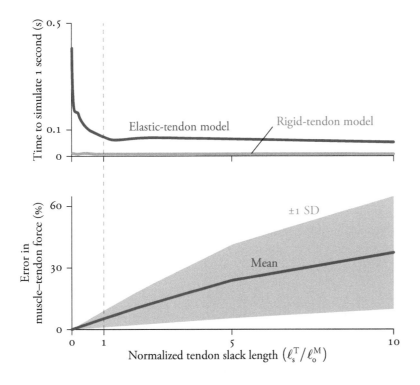

FIGURE 10.8
A rigid-tendon approximation can greatly reduce the time required to simulate a muscle with a relatively short tendon (top), but introduces error in the muscle–tendon force (bottom), particularly when the tendon is long relative to the optimal fiber length. Adapted from Millard et al. (2013).

a model for a research study. For example, one must confirm the suitability of assumptions such as approximating short, stiff tendons as rigid links to improve simulation speed (Figure 10.8). For muscles with short tendons, representing the tendon as an elastic structure dramatically increases the time required to simulate muscle–tendon dynamics with no appreciable improvement in accuracy. However, for muscles with long tendons, errors in muscle fiber lengths (and therefore in muscle forces) are large when tendons are assumed to be rigid. Thus, I typically use a rigid-tendon model to represent muscles with short tendons but include tendon elastic properties when the tendon is at least as long as the muscle fibers.

In 1990, I developed a model of the knee that represents the complex biological contact surfaces with simple geometry (Figure 10.9). Computers were very slow and expensive at the time, so I needed a model that could calculate the moment arms of the muscles crossing the knee with little computational effort. The

FIGURE 10.9
In many studies, the complex biological contact surfaces in the knee (left) can be approximated by simpler geometry (center) and incorporated into a musculoskeletal model (right) to increase simulation speed.

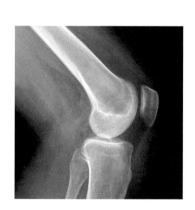
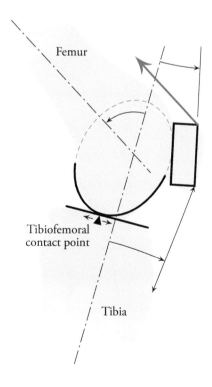

model has continued to perform well in this application for nearly 30 years. However, if I had needed a model to calculate contact stresses within the knee, I would have used more detailed geometric representations of the articular contact surfaces. There are now detailed models of biological joints that exploit the computational resources available today.

We test simulations by comparing quantities that are predicted by the simulation to analogous experimental measurements. As we have already seen, we can compare the joint angles, joint moments, ground reaction forces, and muscle excitation patterns from a muscle-driven simulation to experimentally measured kinematic, kinetic, and EMG data. For example, our simulations of the standing long jump outlined in Chapter 9 included a simple model of the contact between the foot and the ground. Although this model does not reproduce all of the details of experimentally measured ground reaction forces (Figure 10.10), the integral of

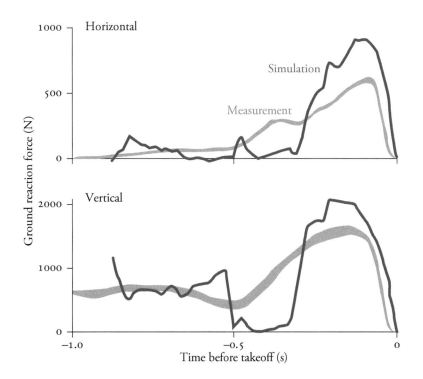

FIGURE 10.10
Horizontal (top) and vertical (bottom) ground reaction forces during the ground contact phase of a standing long jump. The simulation results generated using dynamic optimization (blue) are compared to corresponding experimental data (orange; 95 percent confidence interval, 6 jumps from each of 3 subjects; Ashby and Heegaard, 2002). Adapted from Ong et al. (2016).

our simulated force over time was similar to the integral of the experimentally measured ground reaction force. Because the integral of force is the major determinant of jump distance, we decided that our simple model of foot–ground contact was good enough.

One of the most important and challenging comparisons to make is between simulated muscle activations and experimentally recorded EMG patterns. The comparison is important because the timing of muscle activations influences the interpretation of muscle function. The comparison is challenging because EMG recordings are notoriously noisy and are usually collected for only a subset of muscles. Before Sam Hamner analyzed his simulations of running, he compared his simulated activation patterns to EMG measurements that were collected during his experiments (Figure 10.11). There are no established criteria for determining when the match is good enough, so Sam and I had to use our best judgment to decide how to account for the uncertainty associated with his simulated activations. We found good agreement between the simulated activations and measured EMG once we had considered the electromechanical delay between EMG and force production.

When Kat Steele was a student in my lab, she studied how crouch gait affects the loads in the knee joint. As a first step, she needed to determine whether the loads computed with her model of typical walking matched the knee loads measured by sensors that had been implanted in human knee joints during total knee replacement surgery. When we first made these comparisons, we found that the knee loads predicted by our model were larger than loads we had measured with instrumented implants. We made adjustments to the model until we had sufficiently accurate results (Figure 10.12). We were then confident that we could use the model to study knee loads in crouch gait.

The ability to predict how much metabolic energy is consumed by each muscle during a movement is one of the most powerful features of muscle-driven simulations. It is powerful because we cannot collect these data experimentally, but this presents a challenge for validating calculations of metabolic cost. We typically compare the sum of the energy consumed by all the muscles in the

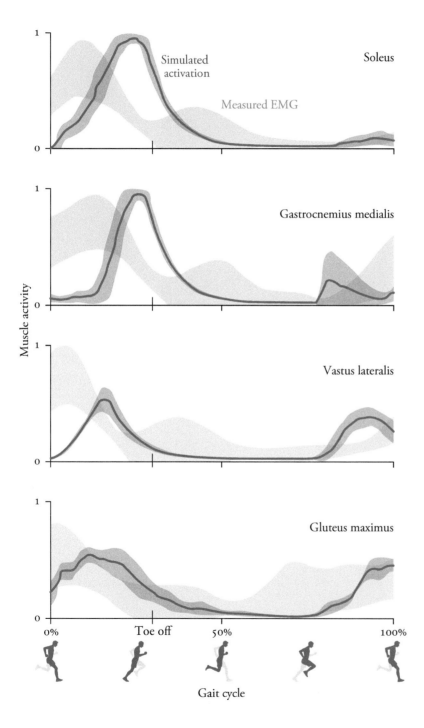

FIGURE 10.11 Simulated activations and EMG recordings of four muscles when running at 5 m/s. The delay of approximately 75 ms between measured EMG and simulated activations is consistent with the electromechanical delay between EMG and force production. Adapted from Hamner and Delp (2013).

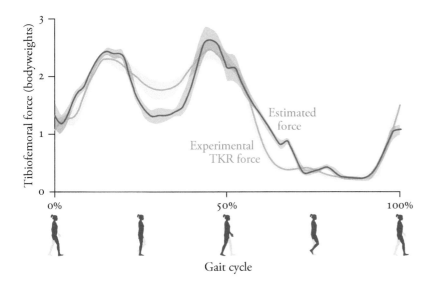

FIGURE 10.12
Tibiofemoral contact force estimated in simulation (blue) and measured by an instrumented total knee replacement (TKR; orange). Mean ±1 standard deviation shown for four trials. Adapted from Steele et al. (2012).

model to estimates of energy expenditure measured via indirect calorimetry (Figure 10.13).

In some studies, we wish to generate muscle-driven simulations for which experimental data would be difficult or impossible to collect. For example, we may wish to evaluate hundreds of assistive device designs without constructing expensive prototypes, recruiting test subjects, or collecting experimental data. In such cases, we can first simulate unassisted walking, for which experimental data are readily available for comparison, and then proceed to investigate assisted walking only once we are confident in our simulation's predictions.

Software tests must also be performed, such as confirming that iterative algorithms converge, numerical integration tolerances are met, and physical laws are obeyed. The process of determining whether a computational model is an accurate representation of the underlying mathematical model is called *verification*. In contrast to validation, where one asks "Am I solving the correct equations?," the verification process involves asking "Am I solving the equations correctly?" We perform hundreds of automated software tests every time we make a change to the OpenSim software to ensure that we are not introducing errors into the code. These tests are

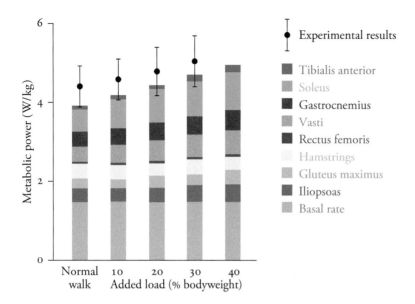

FIGURE 10.13
Average metabolic power consumed during typical walking and when walking while carrying a backpack of increasing mass (up to 40 percent of the subject's body weight). The total metabolic power predicted using dynamic optimization (colored bars) showed trends similar to experimental measurements of total energy consumption (black; mean ±1 standard deviation). The "basal rate" represents the energy consumed while at rest. Adapted from Dorn et al. (2015).

critical because if an algorithm has been implemented incorrectly, all resulting simulations will be wrong and the conclusions of a study will be contaminated. Paradoxically, large errors are least problematic because they are easiest to detect. It is the small errors that can be most pernicious: they can produce results that are plausible and can, therefore, go unnoticed. Verification is facilitated by designing software in a modular fashion, where each component can be tested independently (sometimes called "unit testing") before higher-level tests are performed on a software system.

It is often useful to compute the *sensitivity* of key outputs to parameters of your model that have large uncertainty. To be reliable, the conclusions of a study must be insensitive to small changes in uncertain inputs and arbitrary modeling choices. As we noted in Chapter 8, all models are wrong but some are useful; we must always ensure that simulation results are useful and not merely wrong. As computational biomechanists, we are responsible for ensuring that rigorous software verification has been performed, that our models and simulations have been adequately validated, and that the results we report are insensitive to parameters with large uncertainty.

Stage 4: Analyzing muscle-driven simulations

To gain insight from muscle-driven simulations of human movement, we require more than properly validated simulations. Most importantly, we must pose a question that a simulation can answer (and that an experiment cannot answer more effectively). Simulations compute many quantities that can be measured experimentally, which are useful for validation, but they also provide quantities that cannot be measured in practice. For instance, it is generally not possible to measure the force a muscle generates during a movement; in a simulation, however, muscle forces are computed by a model of muscle contraction dynamics and are readily available for analysis. When combined with calculations of joint kinematics and muscle paths, a model of muscle contraction dynamics allows us to estimate tendon strain, another quantity that is difficult to measure *in vivo* and one that can play a key role in the storage and release of energy during movement.

Simulations allow you to perform powerful analyses. For example, the system of equations governing the dynamics of a model can be solved to determine the contribution of each muscle force to accelerating the model's center of mass (or any other point). This is called a *muscle-induced acceleration analysis*, which uses Equation 10.1 to compute the accelerations of the body that are induced by the application of each muscle force to the skeleton. This analysis applies Newton's second law to determine the accelerations caused by each muscle force, a necessary step to establish cause–effect relationships in a biomechanical system. Note that such an analysis is strictly retrospective and would not be appropriate for predicting new movements or muscle compensation strategies if, for example, a muscle were injured. As we will see in Chapters 11 and 12, a muscle-induced acceleration analysis can reveal how each muscle supports the weight of the body during walking and propels you forward during running.

Simulations also reveal cause–effect relationships, such as how the compliance of the Achilles tendon affects the energetics of running, and can predict the outcome of hypothetical scenarios. Will a model run twice as fast if you double its strength? Will a model adopt a limp or crouched gait when its muscles are weakened? Simulations have begun to answer these questions.

Software for creating muscle-driven simulations

In the early 1990s, Peter Loan and I introduced SIMM: Software for Interactive Musculoskeletal Modeling, a package that allowed users to create, alter, and analyze models of many musculoskeletal structures (Delp and Loan, 1995). Over the next decade, researchers created dozens of models of humans and animals. They used these models to generate simulations of walking, running, cycling, stair climbing, and pathological gait, and studied the biomechanical consequences of surgical reconstructions. SIMM facilitated discovery of principles that govern movement control and treatments for individuals with movement pathologies. It became clear to me that we could accelerate research by expanding the user base and allowing users to share models and simulations more easily.

In 2007, my colleagues and I launched the OpenSim project, an open-source software platform that extends the capabilities of SIMM (Delp et al., 2007). OpenSim's capabilities span three areas. First, users can build, manipulate, and interrogate biomechanical models (Figure 10.14). For example, bone specimens have been used to build a model of the *Australopithecus afarensis* hand to investigate whether this primate species had sufficient grip strength to make certain stone tools (Domalain et al., 2017). Second, OpenSim can enable studies that are difficult to perform experimentally, such as investigating how we exploit tendon elasticity to make running more efficient (Uchida et al., 2016a). Third, using principles of neuromuscular control and dynamic simulation, and without performing any experiments, OpenSim can predict novel movements and adaptations to novel conditions. This capability has led to a deeper understanding of the role that muscles play in preventing ankle injuries (DeMers et al., 2017).

Thousands of biomechanists have shared OpenSim models, simulations, and computational tools on the simtk.org repository,

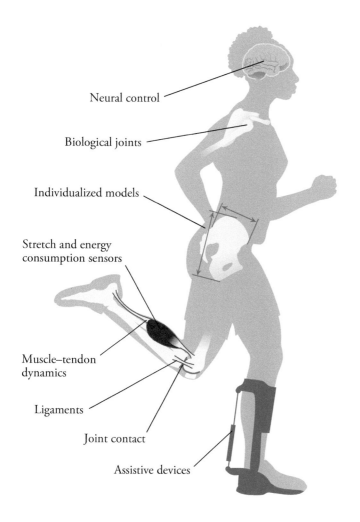

FIGURE 10.14
OpenSim can be used to generate forward dynamic and inverse dynamic simulations of movement. The software allows users to model a wide variety of biological and mechanical systems, including neural controllers, muscle–tendon actuators, and assistive devices. Adapted from Seth, Hicks, Uchida et al. (2018).

which we built to enable researchers to reproduce, validate, and extend the work of others. It is exciting for me to watch the OpenSim community grow and diversify. From running shoes to orthopedic surgeries, the greatest insights are likely to appear through collaboration among experts in a variety of domains. OpenSim provides a platform for this collaboration. In the next two chapters, we will analyze simulations created in OpenSim to study the contributions of muscles to human walking and running, and will see how these simulation results have already been applied in the real world.

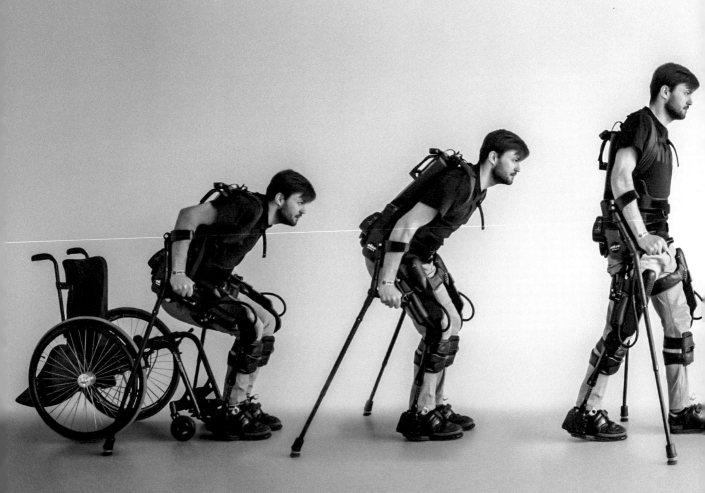

11 Muscle-Driven Walking

We shall not cease from exploration
And the end of all our exploring
Will be to arrive where we started
And know the place for the first time.
—T. S. Eliot

THE VIDEO ENDED and the lights flickered back on. No one knew exactly what to do. The video had shown a 13-year-old girl walking back and forth across our gait lab. The girl had been born six weeks prematurely and was diagnosed with cerebral palsy. She started walking at age 3 and now had a crouch gait. Her knees hurt from the high loads in her joints, and she expended more energy walking on level ground than other kids do running up stairs. She was 20 cm shorter than most of her peers because she walked with flexed hips and knees. My colleagues and I were reviewing her video to devise a strategy to improve her walking, but we could not agree on the best plan of action. In addition to the video, we reviewed a 20-page report from our gait lab, which showed that her joint angles, joint moments, and EMG patterns differed from typical gait, but we still did not know what was causing her crouch gait. Was it tight hamstrings? Weak plantarflexors? Tight hip flexors? If we could identify the underlying causes of her crouch gait, we could suggest a physical therapy program, a leg brace, or an orthopedic surgery to address these causes.

I spent the first 15 years of my career working with other experts to determine the causes of movement abnormalities in children with cerebral palsy. We used all the information we had available and did our best to plan treatments that would improve their ability to walk. There was reason for hope. About half of the children improved substantially following surgery. Unfortunately, many did not, and some got worse. There was a fundamental barrier to developing better treatments: we did not know the contributions of individual muscles to producing joint motions during typical gait or how these contributions differed in children with cerebral palsy. This motivated me to develop tools to help us better understand the actions of muscles during walking.

So far in this book, we have explored the form, function, and simulation of muscle. We now arrive where we started, with the desire to understand walking, but with new tools to deepen our understanding. In Chapter 2, we analyzed walking using models consisting of only a few mechanical linkages representing the legs. Although these simple models are valuable, they do not help us determine the actions of individual muscles. Legs have muscles which generate the forces that keep us from falling down and propel us forward as we walk. Muscles allow us to carry a backpack, change our walking speed, and transition from walking to running.

In this chapter, we will see how muscles are coordinated to produce typical walking and how the dynamics of walking change with speed. We will also see why poorly coordinated muscles result in atypical patterns, such as stiff-knee gait and crouch gait, and what might be done to improve walking dynamics when muscle coordination is impaired.

An important use of muscle-driven simulations is to extend the insights gained from experiments. For example, while it is possible to measure muscle activity, ground reaction forces, and joint motions, experiments alone are insufficient to determine how each muscle contributes to ground reaction forces and body motions. This is the reason we could not determine why the girl in the video was walking with a crouch gait. We identified abnormal muscle activities and abnormal motions, but we could not determine which muscles were causing the abnormal motions. Muscle-driven simulations reveal the motions caused by muscles and provide a powerful tool for understanding muscle actions during movement. In this chapter, we will see how to build, test, and analyze muscle-driven simulations of walking.

Building and testing simulations of walking

To build simulations of walking, we often start by recording the motions of the body segments and ground reaction forces as a person walks over force plates. We may measure other quantities as well, such as muscle activity using EMG or whole-body energy expenditure using indirect calorimetry. We then tailor a generic musculoskeletal model to match the subject's size and shape based on measured positions of anatomical landmarks. We calculate the joint angles during walking using an inverse kinematics algorithm that minimizes the difference between the measured marker positions and the positions of the corresponding markers on the model at each time frame, as described in Chapter 7. We simulate the muscle activations and forces needed to track the measured motions as outlined in Chapter 10, producing a muscle-driven simulation of walking (Figure 11.1). The accuracy of the simulation must then be evaluated as described in detail by Liu et al. (2008) and Hicks et al. (2015). For example, we might compare simulated muscle activations with EMG measurements, or the total metabolic energy consumed by all simulated muscles with measurements of whole-body metabolic cost.

Once the accuracy of a simulation has been tested, we can analyze it to determine how muscles contribute to the vertical and fore–aft accelerations of the body's center of mass, and to the motions of the joints and body segments. Over the last 30 years, my research group has used this approach to analyze thousands of muscle-driven simulations of human gait, including healthy and impaired walking at various speeds. These analyses have allowed us to synthesize a reasonably complete picture of how muscles contribute to support and progression. I summarize our key findings below.

Muscle contributions to ground reaction forces

During the stance phase, muscles generate forces that support the body's weight and regulate forward progression. We can quantify a muscle's contribution to body-weight support by determining how the muscle affects the vertical acceleration of the center of mass. Similarly, we can investigate how a muscle contributes to forward progression by analyzing its role in generating accelerations of the center of mass in the fore–aft direction. Recall that the acceleration of the body's center of mass during walking is related to the ground reaction force divided by the mass of the body (Equation 2.2); thus, analyzing ground reaction forces is akin to analyzing the acceleration of the center of mass. Also recall that the ground

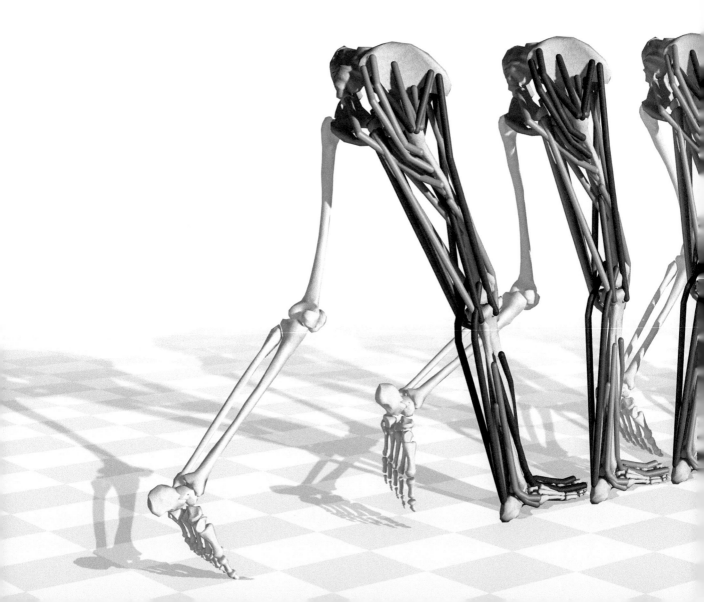

FIGURE 11.1
Visualization of a muscle-driven simulation of walking. The colors of the muscles indicate their level of activation ranging from inactive (blue) to highly active (red). Data from Dembia et al. (2017).

reaction forces are generated by, and can be attributed to, muscle "action" forces.

The same muscles that provide body-weight support during walking also regulate forward progression. For example, a research team at Stanford led by May Liu found that the vasti and gluteus maximus make important contributions to support during early stance while the vasti also reduce the body's forward velocity (Figure 11.2). The gluteus medius provides support during

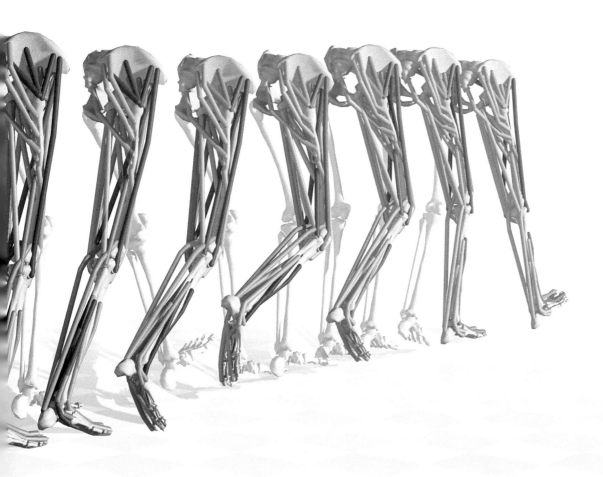

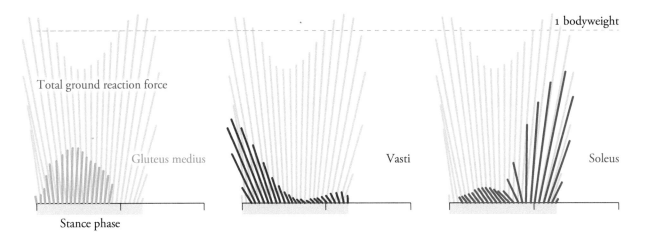
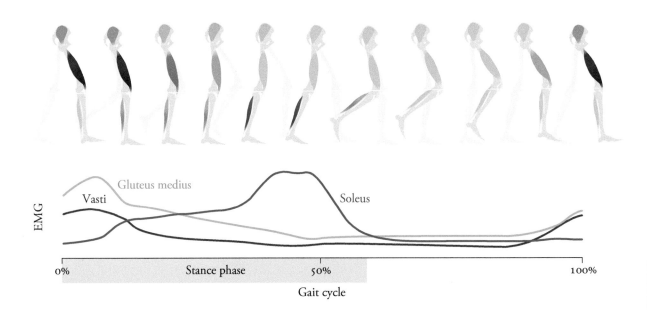

MUSCLE-DRIVEN WALKING 279

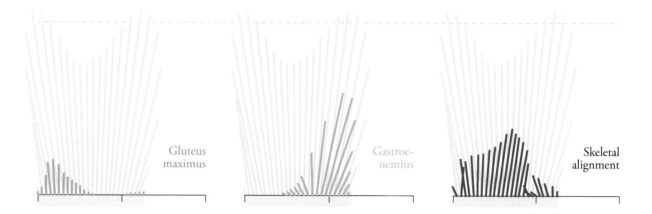

FIGURE 11.2
Actions of the gluteus medius, vasti, soleus, gluteus maximus, and gastrocnemius during the stance phase. The vasti are most active during early stance when they accelerate the center of mass upward and backward; the plantarflexors are active during late stance when they accelerate the center of mass upward and forward. The gluteus medius, gluteus maximus, and skeletal alignment make important contributions to body-weight support. Data from Liu et al. (2008).

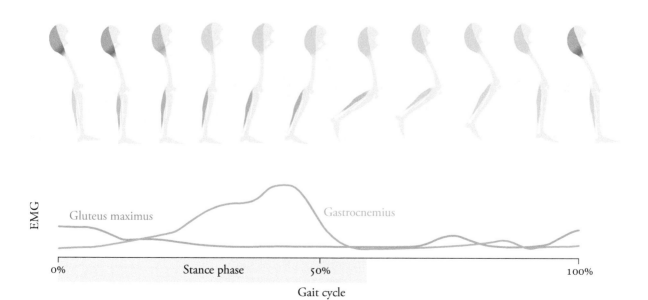

midstance and, in the second half of stance, contributes to forward acceleration. The soleus and gastrocnemius contribute to vertical and forward acceleration in late stance. The contributions to body-weight support from the soleus and gastrocnemius are so important that weakness of these muscles may lead to crouch gait. It is for this reason that, in some cases, crouch gait can be improved by wearing a spring-loaded ankle brace that generates a plantarflexion moment.

For brief periods during stance, the body's center of mass accelerates downward at values close to 9.8 m/s^2. In the absence of muscle forces, the body would be in near-freefall during these periods. When the foot is flat on the ground and the knee is near full extension, however, the magnitude of the mass center's vertical acceleration due to gravity is less than 9.8 m/s^2 because the skeleton provides passive resistance to gravity (see skeletal alignment in Figure 11.2). However, the passive support of the bones is insufficient to prevent buckling of the legs and collapse. Muscles are necessary to support the body's weight during normal walking.

Summing the contributions of the stance-limb muscles to the ground reaction force produces a pattern similar to that of the familiar ground reaction force vectors (Figure 11.3). This indicates that, during walking, muscles are principally responsible for generating the ground reaction force and thus for supporting the body's weight. The difference between the total ground reaction force and the muscle-generated ground reaction force can be attributed to skeletal alignment, or the resistance provided by the skeleton to downward acceleration caused by gravity. During double support, muscles in both limbs contribute to supporting the body's weight; however, the muscles in the trailing limb assist forward progression while those in the leading limb resist it (Figure 2.5).

Muscle actions during the swing phase

The swing phase begins with flexion of the hip, knee, and ankle, which draws the toe of the swing limb away from the ground as the limb moves forward. Flexion of the knee is especially important for

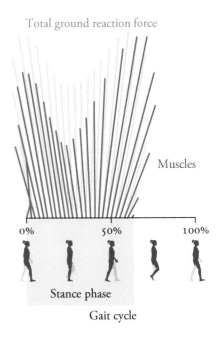

FIGURE 11.3
Contribution to ground reaction force from stance-limb muscles (red) relative to the measured ground reaction force (gray). The difference represents the contribution of skeletal alignment. Adapted from Liu et al. (2008).

toe clearance during swing. Without sufficient knee flexion, the toe of the swing limb will strike the ground. During typical walking, the toe clears the ground by only about 1 cm. With so little margin for error, it's no wonder that we occasionally scuff our feet on the ground.

In the ballistic walking model, the motion of the swing leg is likened to the unforced swinging of a compound pendulum (Figure 2.9). The low level of activity in the leg muscles during swing (relative to stance) supports the characterization of leg swing as being a largely passive motion. However, the leg muscles do exhibit stereotypical patterns of activity before and during the swing phase. Even this relatively small amount of muscle activity plays an important role.

The motion of the swing limb is shaped by muscle forces that are generated before and during swing. In late stance, muscle forces establish the initial conditions for swing. The knee flexion velocity at toe off is a key factor affecting how the limb swings and is determined by the action of muscles even before the foot leaves the ground. Using muscle-driven simulations to analyze the actions of muscles during double support, Saryn Goldberg and her colleagues found that the iliopsoas and gastrocnemius are the largest contributors to increasing knee flexion velocity at toe off; the vasti, soleus, and rectus femoris muscles act to decrease this velocity (Figure 11.4). Thus, the actions of these muscles must be balanced prior to swing so that, at toe off, the right conditions are established for effective swing-phase knee flexion.

Swing dynamics are also influenced by muscle activity during swing. For example, the hamstrings play an important role near the end of the swing phase (Figure 11.5). The hamstrings have a complex action because they cross posterior to the hip and the knee, thereby generating a hip extension moment and a knee flexion moment. The knee flexion moment produces a knee flexion acceleration, but the hip extension moment induces a nearly equivalent knee extension acceleration; thus, the net action of the hamstrings at the knee is small. Allison Arnold analyzed simulations of the swing phase to show that the primary effect of the hamstrings is to

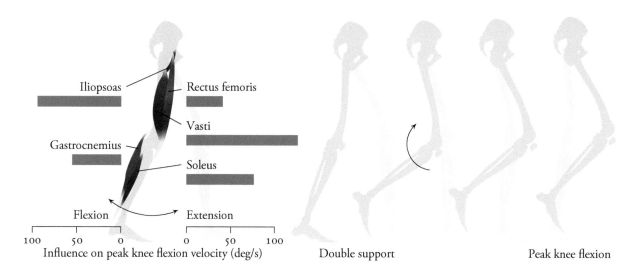

FIGURE 11.4
The knee flexes rapidly during double support, resulting in a high knee flexion velocity at toe off. The knee flexion velocity increases during double support due to forces generated by the iliopsoas and gastrocnemius, and is limited by forces generated by the vasti, soleus, and rectus femoris. Data from Goldberg et al. (2004).

produce a backward acceleration of the shank (Arnold et al., 2007). Thus, the hamstrings slow the limb at the end of swing. The rectus femoris has the opposite action—it gets the limb moving forward at the start of swing.

The rectus femoris muscle also plays a critical role in regulating knee flexion during the swing phase. It has long been thought that excessive activity of the rectus femoris during swing produces a gait abnormality called stiff-knee gait, in which swing-limb knee flexion is reduced and toe clearance is compromised. We have used muscle-driven simulations to investigate the extent to which excessive rectus femoris activity before and during the swing phase contributes to producing stiff-knee gait. We will see the results of this investigation in the next section.

Muscle actions in stiff-knee gait

Stiff-knee gait, characterized by diminished and delayed peak knee flexion in swing (Figure 2.22), is one of the most common gait problems in children with cerebral palsy and in adults who have suffered a stroke. Insufficient knee flexion during swing can lead to tripping and falling as well as inefficient compensatory movements. Despite the prevalence of stiff-knee gait, its causes are not well

FIGURE 11.5
The rectus femoris is active in early swing and accelerates the limb forward. The hamstrings are active in late swing and reduce the forward velocity of the swing leg prior to foot contact.

understood. Several factors may contribute to stiff-knee gait, but excessive activity of the rectus femoris muscle is considered to be a primary cause. Common treatments alter the function of the rectus femoris to improve knee motion during the swing phase. For example, rectus femoris transfer surgery relocates the attachment of the rectus femoris from the patella to a more posterior site (e.g., the sartorius just behind the knee) to augment knee flexion. A nonsurgical treatment option is botulinum toxin injection into the rectus femoris. Botulinum toxin is a neurotoxin that, when injected into a muscle, blocks acetylcholine release at the neuromuscular junction, thereby preventing activation of the muscle.

What are the causes of stiff-knee gait? How important is swing-phase activity of the rectus femoris in producing stiff-knee gait? These questions are important to ask but are not easy to answer. The causes of stiff-knee gait are difficult to determine from experimental data alone and may differ from one individual to another. The outcomes of treatments are also variable—some patients have improved knee flexion after surgery while others do not. These challenges motivated me to develop muscle-driven simulations of stiff-knee gait.

We used muscle-driven simulations to examine how activity of the rectus femoris prior to and during swing contributes to knee flexion. Jeff Reinbolt created simulations that reproduced the dynamics of subjects with stiff-knee gait. To study the effects of rectus femoris activity on knee motion, Jeff eliminated excitation of the rectus femoris in the simulations, first prior to swing and then early in the swing phase (Figure 11.6). We found that knee flexion increased in both cases, but the increase in knee flexion was greater

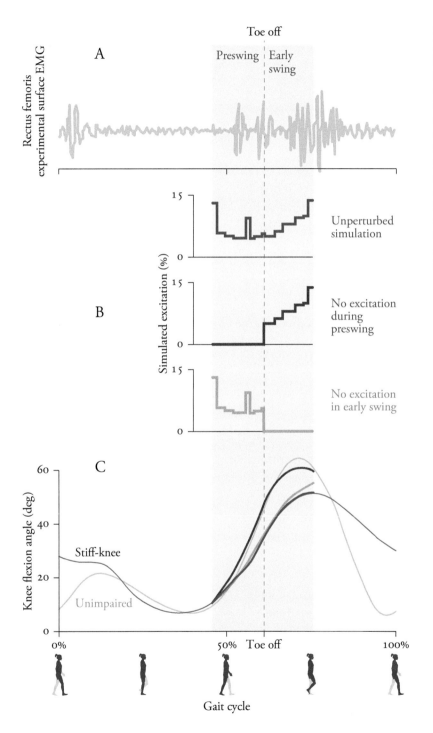

FIGURE 11.6
Methods used to compare increase in peak knee flexion when rectus femoris activity is eliminated during preswing and in early swing. (A) Rectus femoris EMG of a stiff-knee subject over one gait cycle. (B) Eliminating excitation of the rectus femoris during preswing (red line) and in early swing (orange line). (C) Simulated changes in knee flexion angles were larger when rectus femoris activity was eliminated during preswing (red) than in early swing (orange). Adapted from Reinbolt et al. (2008).

when rectus femoris activity was eliminated prior to swing (during "preswing"). This suggests that preswing activity of the rectus femoris may be responsible for limiting the knee flexion of many persons with stiff-knee gait—not activity during swing as is often presumed.

This example illustrates an important characteristic of muscle function—the activity of a muscle affects motions in the future. As we have seen, there is a delay between muscle excitation and muscle force production, as well as a delay between muscle force production and appreciable changes in joint position. Thus, when attempting to identify which muscles are producing an abnormal motion, we must consider muscle activity that precedes the motion. Muscle activity that occurs during or after an event of interest (e.g., at the instant of peak knee flexion) can have no effect on an individual's motion until sometime later, after the development of muscle force. This statement may seem obvious, but it is often forgotten when attempting to determine which muscles cause abnormal motions. I have been in many meetings where an expert suggests that abnormal EMG occurring after an abnormal motion caused the motion. It simply can't happen that way.

My research group has also used muscle-driven simulations of stiff-knee gait to study the mechanisms by which rectus femoris transfer surgery alters muscle function. Melanie Fox altered the rectus femoris in a computer model to simulate the effects of various treatments (Figure 11.7). The simulations showed that the largest improvement in peak knee flexion occurred after simulated transfer of the rectus femoris to the sartorius, a surgery that converts the rectus femoris from a generator of knee extension moments to a generator of knee flexion moments. Transfer of the rectus femoris to the iliotibial band produced a smaller improvement in peak knee flexion. Simulated botulinum toxin injection into the rectus femoris, which eliminates the knee extension moment it generates as well as its hip flexion moment, produced less improvement than either simulated transfer. These simulations suggest that the primary mechanism by which rectus femoris transfer improves knee flexion is the reduction of the

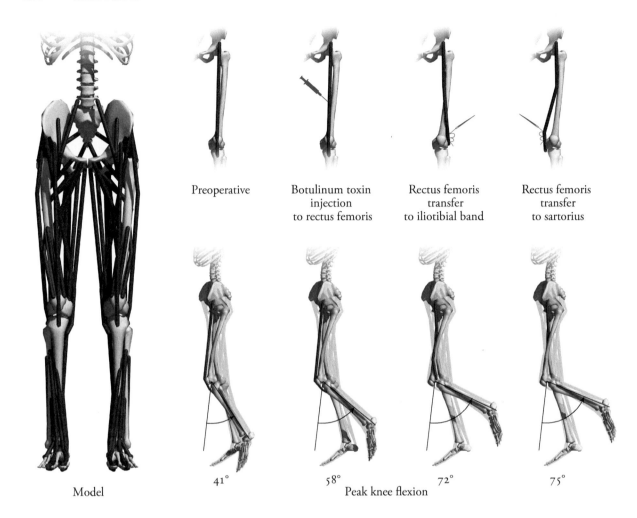

FIGURE 11.7
Peak knee flexion of lower-limb model (left) during simulation of preoperative gait and gait after three candidate interventions (right) for a patient with stiff-knee gait. Adapted from Fox et al. (2009).

muscle's knee extension moment. A secondary mechanism of improvement is preservation of the muscle's hip flexion moment, which induces knee flexion.

We next examined the relationships between knee flexion velocity at toe off, joint moments during double support, and improvements in stiff-knee gait following rectus femoris transfer surgery in individuals with cerebral palsy. Subjects with stiff-knee gait tended to exhibit abnormally low knee flexion velocities at toe off and excessive knee extension moments during double support. Subjects with good outcomes following surgery showed substantial improvement in these metrics postoperatively; subjects with poor outcomes did not. Therefore, although stiff-knee gait is observed during the swing phase, its primary cause is abnormal muscle activity during late stance. As Carl Sagan once said, "You have to know the past to understand the present."

Muscle actions over a range of walking speeds

We saw in Chapter 2 how walking speed influences joint kinematics and ground reaction forces. It should come as no surprise that muscle activations also change with walking speed. Activation patterns of key lower-limb muscles during walking at four speeds are shown in Figure 11.8. To obtain these patterns, Edith Arnold measured surface EMG signals from healthy subjects while they walked on a treadmill. The EMG signals were then high-pass-filtered at 30 Hz, rectified, and low-pass-filtered at 10 Hz (Figure 4.14). Each subject's filtered EMG measurements were normalized by the maximum EMG value detected for each muscle across a range of activities, including walking, running, and maximum voluntary contractions (i.e., after normalization, maximum activity = 1).

During typical walking, we observe activation of the gluteus maximus, gluteus medius, vastus lateralis, and vastus medialis in early stance, all of which make important contributions to the ground reaction force. We also see strong activity of the tibialis anterior in early stance. During typical gait, the tibialis anterior generates force while it lengthens, acting as a brake to control the lowering of the foot to the ground (Figure 11.9); "foot slap" may occur if the tibialis anterior is fatigued or paralyzed. You may have felt soreness in your shins the day after a strenuous hike; this is from the strain the tibialis anterior experienced while serving as a brake. The tibialis anterior is also active during swing, to lift the toes and avoid tripping. In general, muscle activity increases with increasing walking speed (Figure 11.8). In late stance, the gastrocnemius and soleus become highly active as they accelerate the center of mass forward (Figure 11.2). The rectus femoris is most strongly activated at the beginning of the swing phase, whereas the hamstrings

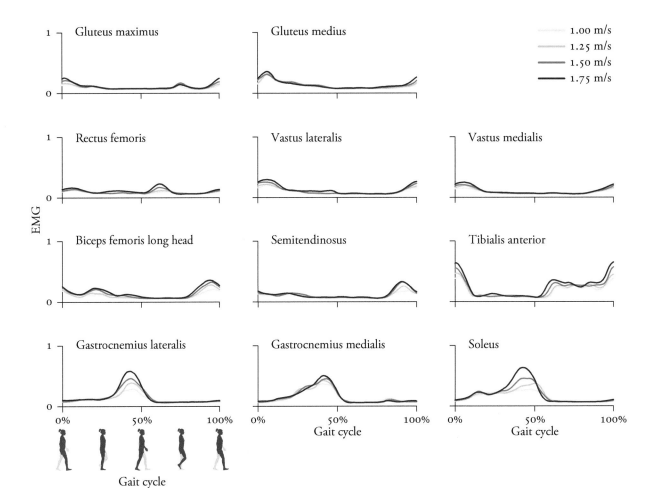

FIGURE 11.8
Average normalized EMG patterns for eleven muscles measured at four walking speeds (1.00–1.75 m/s). The raw EMG signals were rectified, filtered, and normalized to the maximum value detected for each muscle across a variety of activities. Data from Arnold et al. (2013).

(including the biceps femoris long head and semitendinosus) are most strongly activated at the end of swing.

Many individuals with neuromuscular impairments walk slowly. Evaluating a patient's gait requires us to discriminate between deviations caused by pathology and changes that are simply due to walking slowly. Here, we examine how muscles modulate ground reaction forces at various walking speeds. My research group worked with Michael Schwartz, a bioengineer at Gillette Children's Specialty Healthcare, to analyze three-dimensional muscle-driven simulations of walking. We used OpenSim to reproduce the motion of eight healthy individuals walking at four speeds: very slow, slow, natural, and fast (Figure 11.10). These speed categories were defined based on nondimensional walking speeds: "fast" included walking trials that were at least one standard deviation faster than the mean "natural" walking speed for each individual, "slow" trials were between one and three standard deviations slower, and "very slow" trials were more than three standard deviations slower than natural.

We found that muscle contributions to support and progression increase with walking speed, with especially large increases in the vasti between slow-speed and natural-speed walking (Figure 11.11). During very slow and slow walking, a straighter limb in early stance—rather than muscles—provided a majority of the support against gravity.

These results highlight similarities between simple and muscle-driven dynamic walking models. For example, in muscle-driven simulations, we observe forward acceleration (propulsion) from trailing-limb muscles and backward acceleration (braking) from leading-limb muscles during double support, which is consistent with the redirection of the mass center's velocity in simple models. Max Donelan and colleagues (2002) used a simple model with rigid limbs to demonstrate that a force applied to the trailing limb redirects the velocity of the center of mass; muscle-driven simulations of walking reveal that the gastrocnemius and soleus play this role. In simple models, the braking influence of the strut-like leading limb arises from the passive transmission of the ground reaction force to the center of mass (Kuo, 2002). Similarly, in our slowest walking simulations, we observed that skeletal alignment served a strut-like function.

The differences between simple models and muscle-driven models are more apparent at faster walking speeds. As walking speed increases, the magnitude of the fore–aft ground reaction force during early and late stance increases (Figure 2.20), which produces larger braking and propulsive forces. The muscle-driven simulations show that the gastrocnemius and soleus provide the required increased propulsion from the trailing limb. However, due to knee

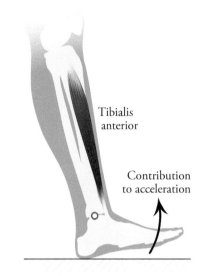

FIGURE 11.9
The tibialis anterior is active at heel strike and generates force to control the lowering of the foot to the ground in early stance.

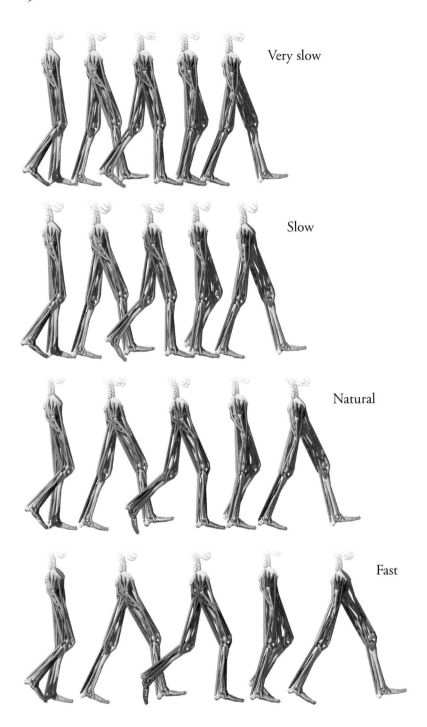

FIGURE 11.10
Visualizations from muscle-driven simulations of a representative subject walking at four speeds. Muscle color indicates simulated activation level, from no activation (blue) to full activation (red). Data from Liu et al. (2008).

FIGURE 11.11 (OPPOSITE)
Contributions to the acceleration of the body's center of mass from key muscles at four walking speeds. Each ray is the resultant vector of the vertical and fore–aft accelerations, averaged across 8 subjects. Adapted from Liu et al. (2008).

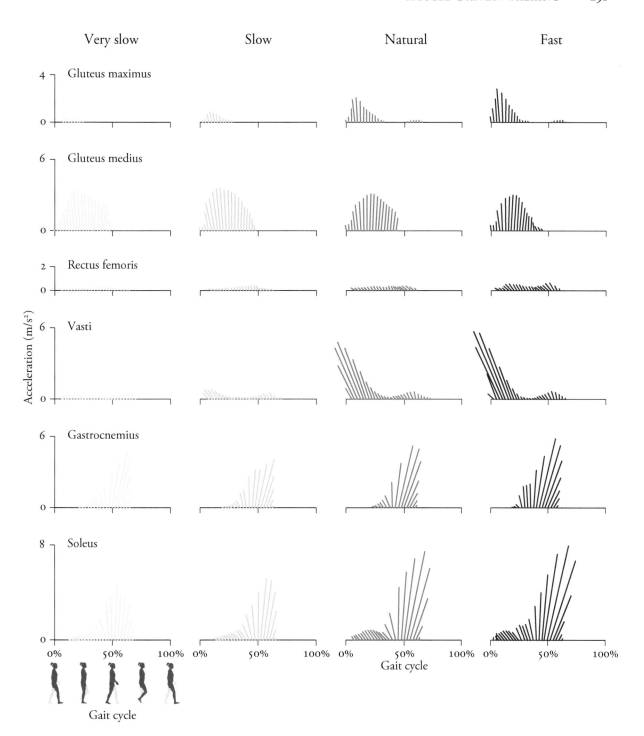

flexion, the leading limb is no longer strut-like at faster walking speeds and cannot transmit the ground reaction forces passively. Knee flexion in humans requires active modulation by the vasti, which provide the braking force from the leading limb.

Muscle actions in crouch gait

Many individuals with cerebral palsy walk in a crouch gait, in which the knee and hip are excessively flexed. Walking in a crouched posture can cause knee pain and increase energy expenditure.

FIGURE 11.12 Musculoskeletal models used to create muscle-driven simulations of individuals with typical gait (left) and individuals with cerebral palsy who walked with mild and moderate crouch gait (center and right). Adapted from Steele et al. (2013).

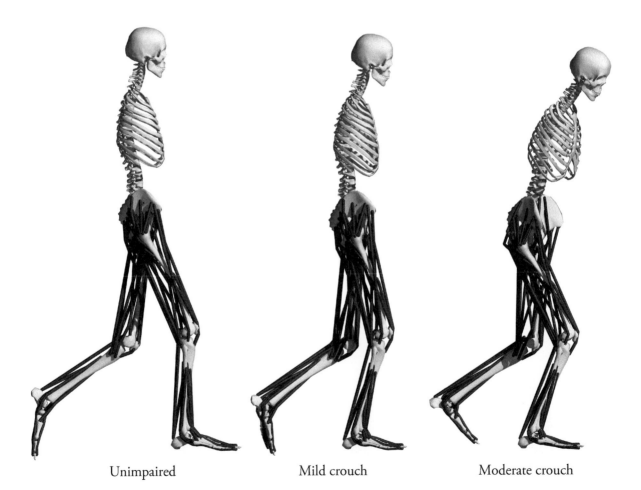

Unimpaired   Mild crouch   Moderate crouch

Understanding how muscles accelerate the center of mass during crouch gait can provide insight into the biomechanics of this condition and suggest methods to improve walking.

Muscle-driven simulations have revealed how the muscle contributions to vertical and fore–aft accelerations of the center of mass differ between crouch gait and typical gait, and how these muscle contributions change with crouch severity. To calculate muscle contributions to the acceleration of the body's center of mass, Kat Steele analyzed three-dimensional simulations of walking for typically developing children and for children with cerebral palsy who walked with varying degrees of crouch severity (Figure 11.12). The simulations revealed that, as in unimpaired gait, two muscle groups are largely responsible for accelerating the center of mass upward and modulating fore–aft acceleration during crouch gait: the quadriceps and the ankle plantarflexors. However, unlike in typical gait, these muscles accelerate the center of mass throughout stance and induce large, opposing fore–aft accelerations. In crouch gait, the ankle plantarflexors accelerate the center of mass upward and forward while the vasti accelerate the center of mass upward and backward throughout stance (Figure 11.13). Proper coordination of quadriceps and plantarflexor activity during unimpaired gait provides an efficient walking strategy. This strategy is compromised during crouch gait, and these muscles are required to make more sustained contributions to vertical and fore–aft accelerations. It's like driving with the brakes on (Figure 11.14).

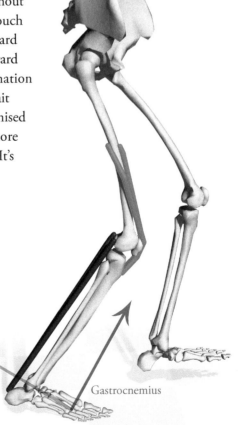

FIGURE 11.13
During crouch gait, the vasti and plantarflexors are active throughout stance and produce sustained ground reaction forces in the upward direction to support body weight, but these muscle groups also produce opposing ground reaction forces in the forward and backward directions.

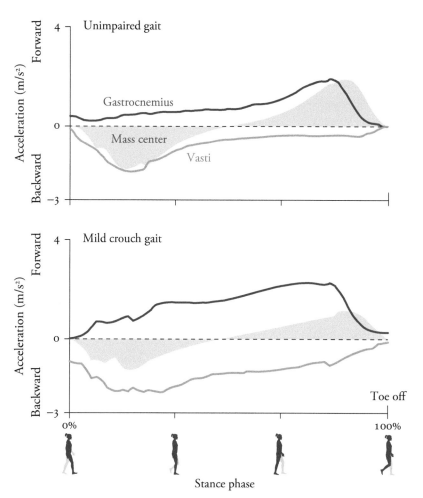

FIGURE 11.14
Fore–aft accelerations of the center of mass produced by the gastrocnemius and vasti during stance. The blue-shaded region indicates the experimentally measured acceleration of the center of mass (i.e., fore–aft ground reaction force divided by body mass). Data from Steele et al. (2013).

In the erect posture of unimpaired gait, a larger portion of body weight is supported by skeletal alignment than when in a crouched posture. As a result, greater forces are generated by the muscles that support body weight in crouch gait, contributing to the inefficiency of walking in a crouched posture. The forces produced by the quadriceps, for example, are significantly greater in crouch gait than in unimpaired gait, and these muscle forces increase with crouch severity (Figure 11.15). Our simulations indicate that increasing crouch severity is also associated with increasing knee loads, which may wear out knee joints and cause pain.

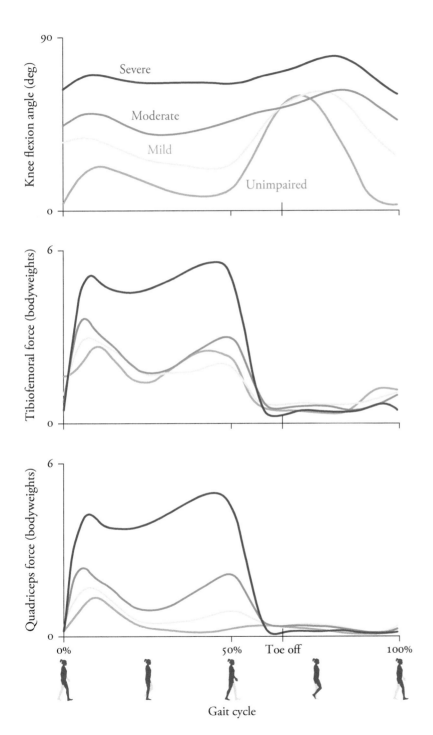

FIGURE 11.15
Average knee flexion angle (top), compressive knee force (center), and quadriceps force (bottom) during one gait cycle for subjects who walked with unimpaired gait and with mild, moderate, and severe crouch gait. Data from Steele et al. (2012).

Our simulations also revealed that a crouched posture increases the demand on muscles to accelerate the center of mass upward due to a decrease in skeletal support. The quadriceps and ankle plantarflexors produce the greatest upward acceleration of the center of mass during crouch gait, but the ability of the gluteus medius to accelerate the center of mass upward is compromised in crouch gait. During unimpaired gait, the gluteus medius accelerates the center of mass upward in midstance during the transition from quadriceps to ankle plantarflexor activity. When walking in a crouched posture, the gluteus medius contribution to upward acceleration is reduced; thus, the contributions from the quadriceps and plantarflexors must be sustained over a greater portion of the gait cycle to support the center of mass in midstance (Figure 11.14).

The above results provide useful guidance on treatment options. Demand on the vasti and rectus femoris is greater when walking in a crouched posture and may contribute to fatigue. Strength training or other programs that improve the endurance of these muscles may help to reduce fatigue in individuals with crouch gait. The ankle plantarflexors are critical muscles for accelerating the center of mass upward and forward during crouch gait and are good targets for strength training programs. Jack Engsberg and colleagues (2006) reported an improvement in knee flexion during stance after strengthening the ankle plantarflexors. Weakening the gastrocnemius or soleus by tendon lengthening surgeries, neuromuscular toxin injections, or other procedures that are commonly performed on individuals with cerebral palsy could *reduce* an individual's ability to support their body weight. Ankle–foot orthoses and active devices that assist the ankle plantarflexors may reduce knee flexion, reduce knee loads, and improve walking efficiency in people with crouch gait.

How did we use insights from muscle-driven simulations to improve the gait of the 13-year-old girl I introduced at the beginning of the chapter? First, we used a musculoskeletal model to determine that her hamstrings were of sufficient length for normal walking, so short hamstrings were not a likely cause of her crouch gait.

She was a candidate for a hamstring lengthening surgery, but I recommended against this surgery. Second, we looked carefully at the strength of her ankle plantarflexors. We understood from the muscle-driven simulations that the soleus muscle is important for generating upward accelerations of the center of mass. Her plantarflexors were weak and could generate less than half of the ankle moments required to walk with a typical gait. Her physical therapist recommended a lightweight, spring-loaded ankle brace that would fit inside her shoe and produce ankle plantarflexion moments to augment her muscles. Two weeks later, she was walking with her new brace in a more upright posture. No painful surgery was necessary. Her plantarflexion moments were greater, and more importantly, the ankle brace allowed her to walk with less flexion of her knees and hips. She still walked with a mild crouch, but she was able to stand taller and keep up with her friends.

Heel-walking and toe-walking

Weakness and tightness of the ankle plantarflexor muscles are often present in individuals with cerebral palsy, stroke, and muscular dystrophy. Recall that muscle contracture (tightness) will result in higher passive stiffness in the affected muscles. Determining cause–effect relationships between these muscular deficits and observed gait pathologies, like heel-walking and toe-walking, is complicated by co-occurring conditions such as skeletal deformities and neural deficits. Carmichael Ong used muscle-driven simulations to predict gait adaptations in response to plantarflexor weakness and contracture, in isolation from other deficits. Carmichael developed a planar model with 9 degrees of freedom (rotation and translation of the pelvis in the sagittal plane, and flexion of the hip, knee, and ankle of each leg) and 9 muscle–tendon actuators per leg representing the major muscle groups of the lower limb. He used dynamic optimization (see Figure 9.15) to find muscle coordination strategies that minimize cost of transport while avoiding falls and injuries. Optimizations using an unimpaired model resulted in gaits that reproduced experimental measurements of kinematics,

Weakness

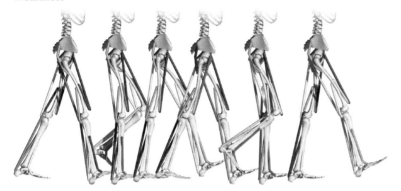

Contracture

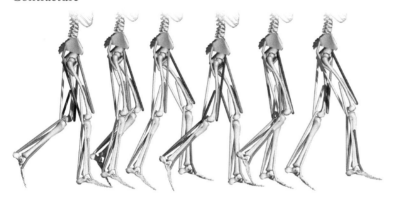

FIGURE 11.16
Dynamic optimizations that minimize cost of transport predict calcaneal gait when plantarflexor muscles are weak (top) and equinus gait when plantarflexor muscles have contracture (bottom). Adapted from Ong et al. (2019).

kinetics, and energy expenditure. After having validated the simulation framework, he repeated the optimization procedure using models with plantarflexor weakness and contracture—under the assumption that these muscular deficits would not affect one's desire to walk with minimum cost of transport. Weakness was modeled by reducing the maximum isometric force of the soleus and gastrocnemius; contracture, by reducing their optimal fiber lengths.

The optimizer produced stable gaits in all cases, regardless of muscular deficit. That is, each model was able to "learn" how to walk. Weakness of the plantarflexor muscles decreased the model's capacity to generate ankle plantarflexion moments, which are critical for producing forward propulsion during walking (Figure 11.2). As a consequence, the weakened model walked slower than the unimpaired model, an effect that is also observed in patients who have plantarflexor weakness. Furthermore, plantarflexor weakness caused the model to adopt a "heel-walking" or calcaneal gait, landing with substantially greater ankle dorsiflexion (Figure 11.16), which is also observed in some patients whose plantarflexors are weak yet flexible. By contrast, plantarflexor muscle contracture resulted in a "toe-walking" or equinus gait, where the model landed on its toes at foot contact and remained on its toes throughout stance. Severe plantarflexor contracture also resulted in crouch gait.

Simulation studies like this one are useful for elucidating cause–effect relationships that would be difficult to disentangle from other factors in an experimental study. This is a critical point because surgical interventions are based on cause–effect reasoning. If we do not know the cause of a patient's crouch gait, or if we assess it incorrectly, we may subject the patient to an unnecessary and ineffective surgery.

Device-assisted walking

Exoskeleton devices can augment the capabilities of the human body. Some of the earliest exoskeleton-like devices were conceived in the late 1800s to assist walking, running, and jumping (Figure 11.17), but only recently have such devices become practical. We can use muscle-driven simulations to improve these devices and guide design of new technologies.

The first powered exoskeleton wasn't built until the 1960s, and it was another four decades before the technology had advanced to the point of practical application. Braces and powered exoskeletons can now improve mobility following a stroke or spinal cord injury (Figure 11.18), support and stabilize the body to prevent injuries, and endow users with superhuman strength. This timeline of about 70 years from conception to successful implementation reflects the considerable challenge of designing, building, and testing

FIGURE 11.17
Illustration from a nineteenth-century patent describing the invention of Nicholas Yagn. From Yagn (1890).

FIGURE 11.18
One of the first exoskeletons to help patients with stroke and spinal cord injury regain bipedal locomotion. Image courtesy of Ekso Bionics.

devices that interact harmoniously with a complex, finely tuned biological system.

A common goal among device designers has been to reduce the metabolic cost of unimpaired walking—a challenging task given the efficiency of natural gait. Designers have made progress using physical prototypes and relatively simple mathematical models. The first exoskeleton to reduce the metabolic cost of walking was reported by Philippe Malcolm and colleagues in 2013. Their device was worn on the shank and used a pneumatic muscle that was attached to the shoe (and tethered to an air compressor) to apply an ankle plantarflexion moment during late stance. Malcolm reported a metabolic savings of about 6 percent compared to natural walking.

The following year, Luke Mooney and colleagues reported savings of about 10 percent with a powered but untethered device that used a similar assistance strategy. In 2015, Steve Collins and colleagues accomplished the same feat with an unpowered ankle exoskeleton, leading to the provocative suggestion that the structure of the human body could, in theory, further evolve to be more efficient during walking. Their device reduced metabolic cost by about 7 percent using a passive clutch mechanism to engage and disengage a spring acting in parallel with the calf muscles. This assistance strategy exploits the fact that, unlike a muscle, a clutched cord requires no energy to maintain its length under tension.

Despite these amazing accomplishments, experimental approaches to device design are limited in several ways. For example, building physical prototypes can be expensive and time-consuming, and it can be difficult to obtain insight about interactions between the musculoskeletal system and an assistive device from experiments alone—that is, without knowledge of muscle forces or muscle energy consumption.

Muscle-driven simulations can be used to explore new assistance strategies without building physical prototypes. Simulations are also useful for testing hypothetical, ideal devices—devices that are massless, provide lossless transmission of torques to the limbs, and have no torque or power limits. Thus, we can use simulations of ideal devices to compare assistance strategies independent of other practical challenges. For example, Chris Dembia and colleagues generated muscle-driven simulations of walking while carrying a 38 kg load, and predicted the effectiveness of seven ideal assistive devices for load-carrying tasks (Figure 11.19). Although devices that apply ankle plantarflexion torques have received much of the focus to date, other assistance strategies (e.g., assisting hip flexion and abduction) may provide greater metabolic savings while also requiring less actuator power.

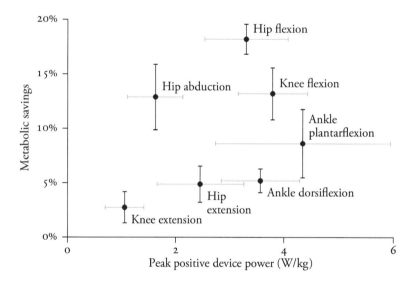

FIGURE 11.19
Muscle-driven simulations predicted the performance of seven ideal devices for assisting walking while carrying a 38 kg load. The most promising devices are those that achieve high savings but have low power requirements. Adapted from Dembia et al. (2017).

Muscle-driven simulations complement experiments by providing detailed analyses of muscle-level dynamics and energy consumption. Chris found that, in general, the optimal device torque profiles differ from the net joint moments generated by muscles during unassisted gait. For example, the ideal ankle plantarflexion device applied a torque that peaked later in the gait cycle and had only about half the magnitude of the ankle moment generated by muscles when unassisted (Figure 11.20). The ankle plantarflexion device did most of the work performed by the soleus, but the gastrocnemius remained active due to its contribution to the required knee flexion moment. Assistive devices can affect the activity of muscles throughout the limb; muscle-driven simulations can provide biomechanical explanations for these effects, and can help determine the timing and magnitude of optimal assistive torques. Computer-aided design has already revolutionized the process of engineering products from appliances to airplanes; it has similar potential for accelerating development of exoskeletons.

FIGURE 11.20
Muscle-level analysis of ankle plantarflexion device for assisting loaded walking (average from 7 subjects). Our simulations predicted differences between joint moments when unassisted (dashed lines) and assisted (solid lines), with far greater reductions in the soleus than the gastrocnemius. Adapted from Dembia et al. (2017).

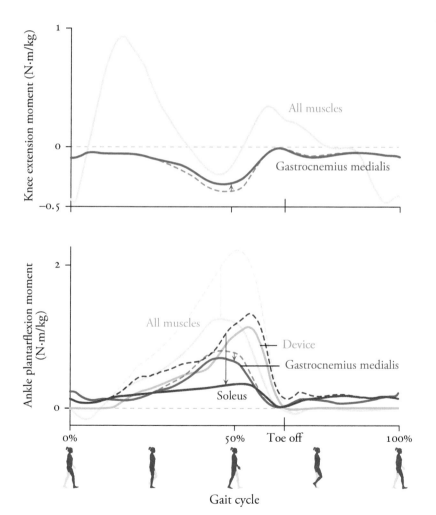

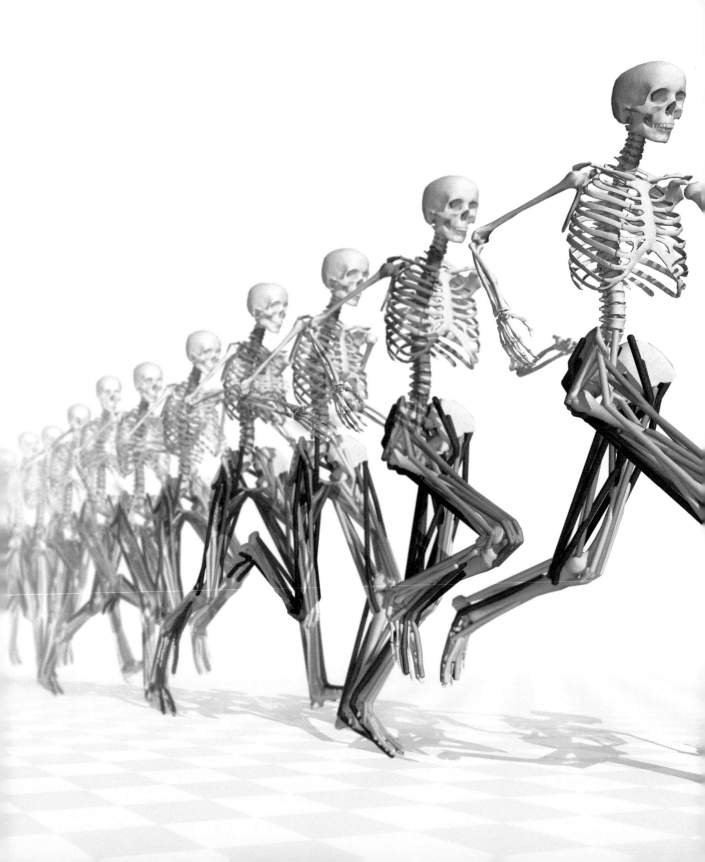

12 Muscle-Driven Running

> I let my feet spend as little time
> on the ground as possible.
> —Jesse Owens

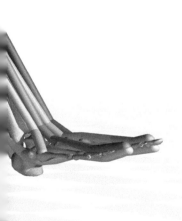

OUR HUMAN ANCESTORS started running as a form of transportation over 2 million years ago. Today, we run for exercise, when we're in a rush, and just for the love of it. We are not the fastest runners on the planet—that distinction goes to the cheetah with a top speed of about 30 m/s, well over twice as fast as the best human sprinters. Nevertheless, our bodies, and those of our ancestors, include features that make us excellent distance runners.

Our distance-running ability is tested each year in the Man versus Horse Marathon. The races began when Gordon Green, a Welsh pub owner, got into an argument with a friend about whether horses or humans were better runners. Green believed that a human could beat a horse over long distances and initiated the Marathon in 1980 to settle the disagreement. The race distance was 22 miles, slightly shorter than an official marathon. Riders on horseback defeated human runners for over two decades. But in 2004, a British man named Huw Lobb finished the race in 2 hours, 5 minutes and 19 seconds, beating the fastest horse by about 2 minutes. Three years later, two other men outran their quadrupedal competitors.

One of the most important features of human running that enables us to move efficiently over long distances is our bouncing gait. In running, the body's center of mass slows and lowers during the first half of the stance phase, storing elastic energy in muscles and tendons. The stored energy is then liberated from these tissues during the second half of the stance phase to accelerate the center of mass forward and upward into flight. This remarkable feat takes place with each step during the 200–300 ms when the foot is in contact with the ground.

Analyses of mass center trajectories and ground reaction forces have suggested that the storage and release of elastic energy is important for running efficiently (Cavagna et al., 1976). This finding motivated researchers to develop models of running in which all lower-limb muscles are represented by a single spring (Figure 3.5). These simple mass–spring models have provided valuable insight into the dynamics of running, and enabled researchers to tune a track to increase running speed (Figure 3.10). While mass–spring models provide a valuable theoretical framework for examining running dynamics, these models are not intended to describe the actions of individual muscles. Yet examining muscle actions during running is essential to understand how muscle architecture and tendon compliance affect running performance, and to inform the design of training programs and technologies that increase running speed and efficiency.

Muscle-driven simulations of running provide this deeper view. They complement the mass–spring models we discussed in Chapter 3 by providing a systematic methodology for estimating muscle forces and their contributions to ground reaction forces. We will see in this chapter that the mass–spring model tells only part of the story. Yes, humans run efficiently in part because the Achilles tendon stores elastic energy. But the elasticity of this tendon also allows the attached muscles to operate in a more efficient regime. Counterintuitively, when we increase our pace and transition from a fast walk to a slow run, the contractions of our plantarflexor muscles become *slower*, allowing the muscles to generate force more efficiently. This phenomenon could not have been predicted with a mass–spring model.

This chapter examines how muscles contribute to the ground reaction force and thus to the support and propulsion of the body's center of mass during running. We will study what the muscles are doing at a typical long-distance running pace of 4 m/s and describe how muscle activations, ground reaction forces, and the muscles' contributions to support and propulsion change with running speed—from a speed that is just above the walk-to-run transition pace of 2 m/s to a brisk run at 5 m/s. We will also see how sprinting is different from lower-speed running and will examine a topic that has been controversial in the running community: the differences between forefoot and rearfoot striking. Finally, we will study how technology can improve running performance.

Building and testing simulations of running

To study muscle actions during running, Sam Hamner measured body segment motions, ground reaction forces, and electromyographic (EMG) patterns from key muscles as experienced runners ran on a treadmill at 2, 3, 4, and 5 m/s (Hamner and Delp, 2013). We used these data to create muscle-driven simulations of each person running at each speed using OpenSim. To do this, we first scaled a generic dynamic musculoskeletal model (Figure 12.1) to match each subject's size, based on the measured locations of markers placed on anatomical landmarks. We calculated the joint angles over the gait cycle using an inverse kinematics algorithm that minimized the difference between the experimentally measured marker positions and the positions of the corresponding markers attached to the model, as described in Chapter 7. We calculated the joint moments using inverse dynamics methods (Chapter 8) and then estimated the muscle activations needed to generate those moments using the

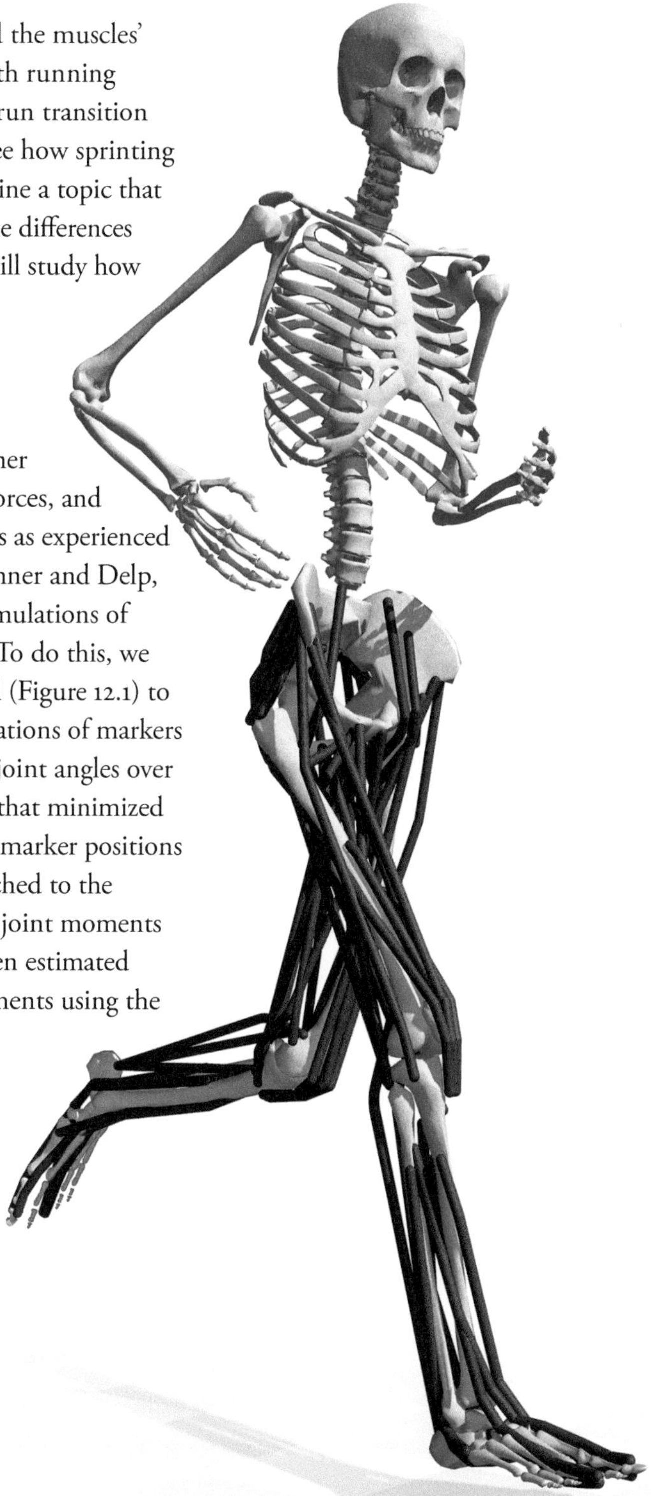

FIGURE 12.1
Generic musculoskeletal model used to generate muscle-driven simulations of running. The model's 12 segments and 29 degrees of freedom are actuated by 86 lower-limb muscles and by torque actuators in the upper body (Rajagopal et al., 2016).

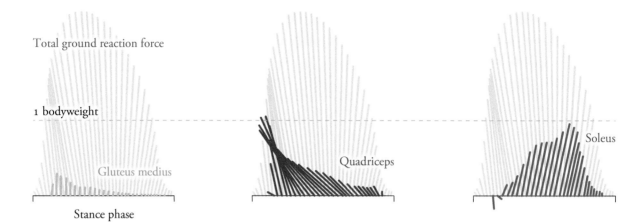

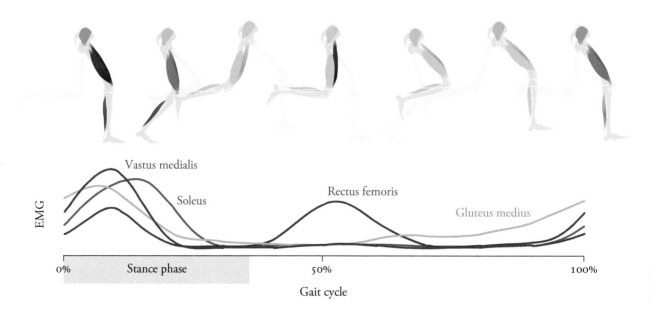

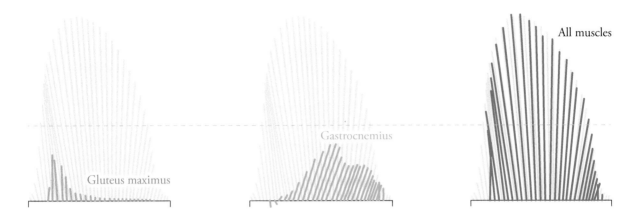

FIGURE 12.2
Actions of the gluteus medius, quadriceps, soleus, gluteus maximus, and gastrocnemius during the stance phase of running at 4 m/s. The quadriceps are most active during early stance when they accelerate the center of mass upward and backward; the gastrocnemius and soleus are active throughout stance, accelerating the center of mass upward and forward. The gluteus medius and gluteus maximus also make important contributions to body-weight support. Adapted from Hamner et al. (2010).

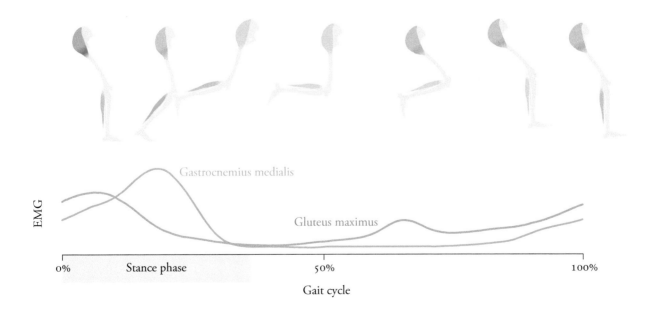

computed muscle control algorithm (Chapter 10). We assessed the accuracy of the simulations through comparison to experimental data, including the body segment motions and EMG recordings, as described in Chapter 10. Finally, we used an induced acceleration analysis to determine the contribution of each muscle force to generating the measured vertical and fore–aft ground reaction forces.

Muscle contributions to ground reaction forces

In running, the center of mass is accelerated by gravity and the muscles of the stance limb (Figure 12.2). During early stance, the quadriceps are the largest contributors to upward and backward accelerations of the center of mass. Throughout most of stance, the ankle plantarflexors (soleus and gastrocnemius) are the main contributors to upward and forward accelerations of the center of mass. The gluteus maximus makes substantial contributions to supporting the body's weight. There are, of course, dozens of other muscles generating forces that influence the motions of the joints and body segments. You can determine the actions of each muscle through a detailed analysis of a muscle-driven simulation; our simulations are freely available at simtk.org.

The soleus is the largest contributor to upward and forward mass center accelerations because it produces large forces and crosses only the ankle. The gastrocnemius, the other major ankle plantarflexor, crosses the ankle and the knee; the knee flexion moment it generates reduces its capacity to accelerate the center of mass upward. Skeletal alignment makes almost no contribution to acceleration of the mass center during running. In slow walking, the lower limb is nearly straight and the skeleton provides passive support, but this is not the case in running. Muscle forces produce nearly all of the ground reaction force (top right panel in Figure 12.2).

Muscle activity during running

Figure 12.3 shows EMG recordings from 11 lower-limb muscles during running at four speeds. These data have been filtered to eliminate high-frequency and low-frequency noise. Each subject's EMG measurements were then normalized to the maximum value recorded for each muscle across all speeds to obtain a value between 0 and 1.

We observe relatively high activation of the gluteus maximus, gluteus medius, rectus femoris, vasti, gastrocnemii, and soleus during the stance phase. The rectus femoris is most strongly activated at the beginning of the swing phase at high running speeds; this muscle crosses in front of the hip and knee, and accelerates the leg forward at the beginning of swing. The hamstrings (biceps femoris long head and semitendinosus) are

FIGURE 12.3
Average EMG from 11 muscles during running. Adapted from Arnold et al. (2013).

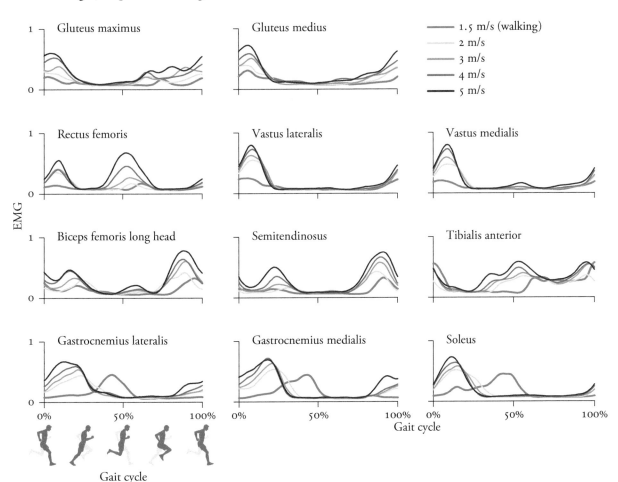

most strongly activated at the end of swing; these muscles cross behind the hip and knee, and accelerate the leg backward at the end of swing. The tibialis anterior is most active during the swing phase to lift the foot. As speed increases, muscle activity increases, and the plantarflexors are recruited earlier in the gait cycle as the stance phase shortens.

Changes with running speed

As one might expect, an increase in running speed is accompanied by increases both in muscle activations and in the upward and fore–aft mass center accelerations produced by the gluteus maximus, gluteus medius, rectus femoris, vasti, and gastrocnemii. The changes in activation, as measured by EMG signals, are shown in Figure 12.3. Note, in particular, that the activation patterns change dramatically as we move from a brisk walk to a slow run. The difference between a slow and a fast run is mostly one of magnitude, with muscles like the rectus femoris and biceps femoris long head increasing their activity dramatically with increasing speed. Though the increase in the activity of the soleus is not as dramatic, it is particularly important. As speed increases, the contribution to upward and forward mass center accelerations by the soleus also increases. The soleus plays a key role in increasing flight times and stride lengths (Figure 12.4; Table 12.1), thereby providing a mechanism for increasing running speed (Equation 3.1).

When running at 5 m/s, soleus muscle fibers shorten with the highest velocities of any muscle during late stance, which coincides with increased ankle plantarflexion speed and decreased stance time compared to lower-speed running. Because runners generate larger upward and forward mass center accelerations at faster running speeds (i.e., ground reaction forces are higher), and because the soleus is the largest contributor to these accelerations, limited force production from the soleus due to its high shortening velocity may put a cap on a runner's top speed. Stated positively, a strong soleus muscle enables runners to generate the large ground reaction forces they need to increase stride length and running speed.

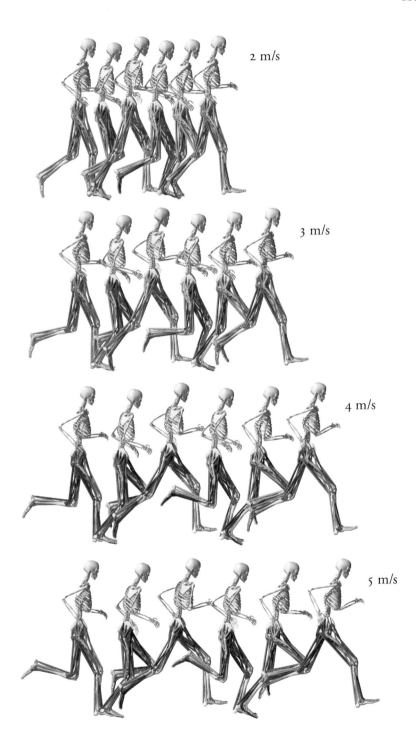

FIGURE 12.4
Visualizations from muscle-driven simulations of a representative subject running at four speeds. Muscle color indicates simulated activation level, from no activation (blue) to full activation (red). Adapted from Hamner and Delp (2013).

Speed (m/s)	Toe-off timing (percent gait cycle)	Stride length (m)	Stride time (s)	Stance time (s)
2.0	47	1.5	0.75	0.34
3.0	40	2.1	0.71	0.29
4.0	38	2.7	0.67	0.26
5.0	37	3.0	0.62	0.24
7.0*	25	4.0	0.57	0.15
9.0*	25	4.1	0.46	0.12

TABLE 12.1
Variation of running parameters with speed: toe-off timing, stride length, stride time, and stance time, averaged across 10 subjects. Data from Hamner and Delp (2013); *Dorn et al. (2012).

Run-to-sprint transition

The changes in joint angles we observed for running speeds up to 5 m/s continue up to about 7 m/s. In general, maximum joint angles continue to increase with increasing speed, as we saw in Figure 3.20. Below about 7 m/s, runners increase their speed primarily by generating greater ground reaction forces, substantially increasing stride length while increasing stride frequency only slightly (Figure 12.5). The increases in ground reaction forces are the consequence of increases in muscle "action" forces, particularly in the ankle plantarflexors.

The muscles must generate larger ground reaction forces to run faster, and they must generate these forces in less time. (As Jesse Owens observed in this chapter's epigraph, he ran fast by touching the ground for as little time as possible.) Thus, the plantarflexors, which are key muscles in generating the ground reaction force, shorten more quickly as running speed increases. Recall that increasing a muscle's shortening velocity decreases its force-generating capacity (Figure 4.9). Beyond speeds of about 7 m/s, the ankle plantarflexors are contracting so quickly that they cannot generate enough force to continue increasing stride length.

At very high running speeds, rather than push on the ground with more force, the strategy for increasing speed switches to pushing on the ground more frequently. Stride length reaches a maximum when running at around 7 m/s; higher running speeds are achieved primarily by increasing stride frequency. The hip

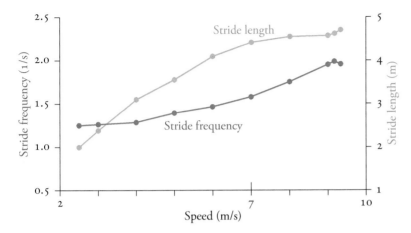

FIGURE 12.5
Increases in running speed are achieved primarily by increasing stride length at low speeds and increasing stride frequency at high speeds. Data from one representative subject are shown. Adapted from Weyand et al. (2000).

muscles, primarily the iliopsoas, rectus femoris, gluteus maximus, and hamstrings, accelerate the limb forward and backward vigorously during the swing phase of high-speed sprinting to swing the legs rapidly. Thus, the highest running speeds are achieved by generating large ground reaction forces during stance, which produces a long stride length, as well as swinging the legs rapidly. Stated another way, our maximum running speed is limited by the ability of our plantarflexors to generate high force at high speed and by the rate at which our hip and knee muscles can swing our legs. If you jump back to Figure 3.7, you will notice that kangaroos have similar trends in stride length and stride frequency as hopping speed increases.

Muscle actions during the walk-to-run transition

Humans transition from walking to running at around 2 m/s, the speed at which running becomes more energetically economical (Figure 3.18). The majority of the metabolic cost of locomotion is consumed by muscles, so we might expect some muscles to become more efficient when making the transition from fast walking to slow running. In fact, an interesting mechanism can be observed in the soleus and gastrocnemius, and their interaction with the Achilles tendon, at the walk-to-run transition. As we walk faster, the shortening velocities of our ankle plantarflexors increase and therefore they produce less force for a given level of activation. The transition from walking to running increases the ability of these muscles to generate force by reducing their fiber velocities (Figure 12.6). Muscle-driven simulations reveal a significant difference in f^V, the effect of fiber velocity on muscle force, between fast walking and slow running in the gastrocnemius lateralis and soleus. Even though gait speed and the shortening velocities of the muscle–tendon units are higher in running, the transition from walking to running decreases the shortening velocities of the muscle fibers, moving gastrocnemius lateralis from concentric to eccentric contraction and soleus from concentric to isometric contraction. Thus, the transition from walking to running is characterized by a *slowing* of the plantarflexor muscle fibers. This striking result, supported by ultrasound measurements (Farris and Sawicki, 2012), occurs because the stretch of the Achilles tendon is greater in running, which better positions the muscles to generate force with less energy.

FIGURE 12.6
Mean and one standard deviation (n = 5) of the force–velocity multiplier, f^V, at the instant of largest active force in walking at 1.75 m/s (blue dot) and running at 2.0 m/s (red dot), plotted on the force–velocity curve. The switch from walking to running decreases fiber shortening velocity and increases f^V in gastrocnemius lateralis and soleus. Data from Arnold et al. (2013).

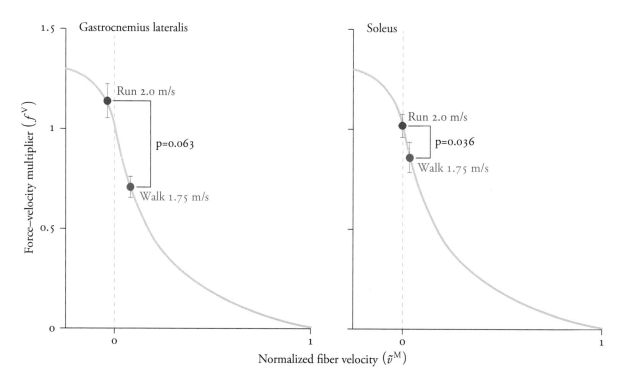

We rely on the storage and release of elastic energy as our muscles and tendons stretch and recoil. Because muscles attach to bones via tendons, the interaction between our muscle fibers and tendons plays a critical role in determining running economy. Lai et al. (2014) found that the elastic strain energy of the Achilles tendon accounts for a large proportion of positive muscle–tendon work, and that this proportion increases with running speed. As we saw above, the compliance of the Achilles tendon reduces the shortening velocity of the soleus and gastrocnemius muscle fibers during running, allowing them to operate at lengths and velocities at which they generate force efficiently. Compliant tendons also store and release energy during running to stretch our limited energetic budget, similar to regenerative braking in electric cars. At the end of this chapter, we will see a device that uses the storage and release of elastic energy to reduce the metabolic cost of running.

Interaction of arm and leg dynamics

Muscle-driven simulations reveal that the arms do not contribute substantially to either propulsion or support when running at or below 5 m/s. However, the angular momentum of the arms about a vertical axis passing through the center of mass counterbalances (i.e., is equal and opposite to) the angular momentum of the legs about the vertical axis (Figure 12.7). Running speed has a significant effect on peak vertical angular momentum of the arms and legs, each doubling from about 1.5 kg·m^2/s at 2 m/s to 3.0 kg·m^2/s at 5 m/s.

Swinging the legs in running

In the early 2000s, Jesse Modica and Rodger Kram sought to understand the role muscles played in swinging the legs during running and the amount of metabolic energy that was consumed in the process. They recorded EMG from key muscles and measured whole-body metabolic cost during moderate-speed treadmill running. Based on EMG timing alone, one may surmise that the rectus femoris and hamstrings are responsible for initiating and

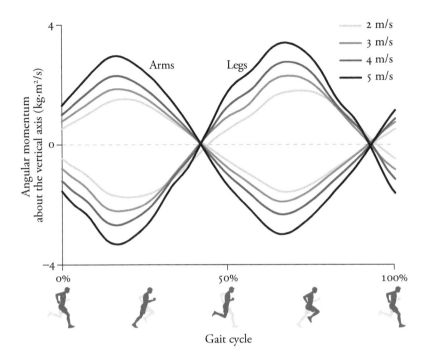

FIGURE 12.7
The opposing angular momenta of the arms and legs about a vertical axis passing through the body's center of mass during running, averaged over 10 subjects. Adapted from Hamner and Delp (2013).

arresting leg swing. To test this hypothesis, Modica and Kram constructed an ingenious apparatus that applied a forward force to each foot during only the first part of the swing phase (Figure 12.8). During assisted running, they observed significant decreases in the activity of the rectus femoris in the first half of swing but observed no decreases in gastrocnemius or soleus EMG, suggesting that the rectus femoris plays a prominent role in leg swing during natural (unassisted) running and the ankle plantarflexors do not. They also observed a decrease in average net metabolic rate; this and follow-on studies suggest that leg swing accounts for approximately 10 to 20 percent of the energy consumed during moderate-speed running. The rest of the energy is consumed to support the body's weight and propel it forward.

Foot-strike patterns

In Chapter 11, we saw that "toe-walking" is observed in people with severe tightness of the ankle plantarflexors. In people with

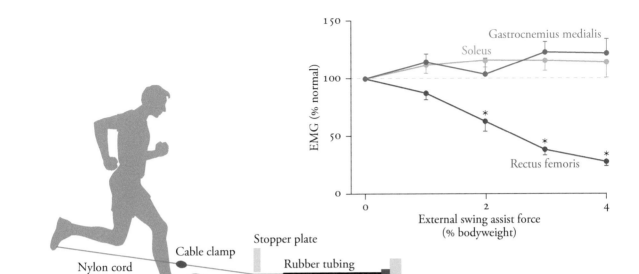

FIGURE 12.8
Swing-assist mechanism and effect on EMG during treadmill running. Left: The rubber tubing stretched as the foot moved posteriorly during stance, applied anterior forces to the foot at the beginning of swing, and became slack when the cable clamp met the stopper plate. Right: EMG magnitudes during running with increasing swing assistance, averaged over the periods during which each muscle is active when running without assistance (n = 10, mean ± standard error; *$p < 0.0001$). Adapted from Modica and Kram (2005).

healthy muscles, toe-walking is rare (though it is occasionally used over short distances for sneaking up on people). By contrast, many healthy runners are natural forefoot strikers, who land on the balls of their feet rather than their heels (Figure 12.9). Running coaches debate whether forefoot or rearfoot striking is best. The vast majority of long-distance runners are rearfoot strikers, but neither foot-strike pattern has been shown to be more efficient.

The kinematics and kinetics of these foot-strike patterns are noticeably different. We saw in Chapter 3 that the vertical ground reaction force rises more gradually in forefoot strikers than in rearfoot strikers (compare Figures 3.2 and 3.3), which has led people to believe that forefoot striking may help prevent bone stress fractures. However, the strain rate in the Achilles tendon is higher when forefoot striking (Lyght et al., 2016), indicating that forefoot striking may have a higher risk of tendon injury than rearfoot striking. Indeed, habitual rearfoot strikers report calf muscle soreness when they convert to forefoot striking, which suggests that

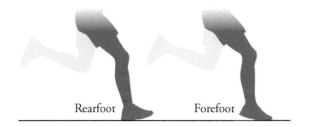

FIGURE 12.9
Foot-strike patterns in running.

the plantarflexors and the Achilles tendon may be loaded differently in the two running styles.

To examine the differences in plantarflexor muscle–tendon mechanics between forefoot and rearfoot striking, Jenny Yong and others in my laboratory generated simulations using data from 16 healthy runners. All the runners were habitual rearfoot strikers, and one set of simulations was generated from data collected while they ran at their natural speed and using their habitual rearfoot-striking gait. The runners then underwent a training protocol whereby they learned to adopt a forefoot-striking pattern. A second set of simulations was generated from data collected during running while forefoot striking. We measured kinematics, which were used to prescribe the joint angles of a muscle-driven model, and EMG, which was processed and applied to the simulated muscles as excitations. The simulations were used to compute muscle activations, fiber lengths, fiber velocities, and the capacity of each muscle to generate active force (Figure 12.10).

Our simulations show that foot-strike pattern affects the gastrocnemius and soleus muscles differently. When forefoot striking, the activation of the gastrocnemius is greater during the first part of stance, resulting in greater forces in the gastrocnemius fibers and the corresponding component of the Achilles tendon. By contrast, the activation of the soleus is lower during push off (at about 25 percent of the gait cycle) when forefoot striking.

There are also similarities in how foot-strike pattern affects the gastrocnemius and soleus. Both muscles have shorter fibers from late swing to early stance during forefoot striking, and fiber velocities are lower in late swing but higher in early stance. During much of early

FIGURE 12.10 (OPPOSITE)
Muscle activation and fiber dynamics of plantarflexor muscles from simulations of running at 2.94 ± 0.30 m/s while rearfoot and forefoot striking (n = 16, mean ±1 standard deviation; *$p < 0.05$). Data from Yong et al. (2020).

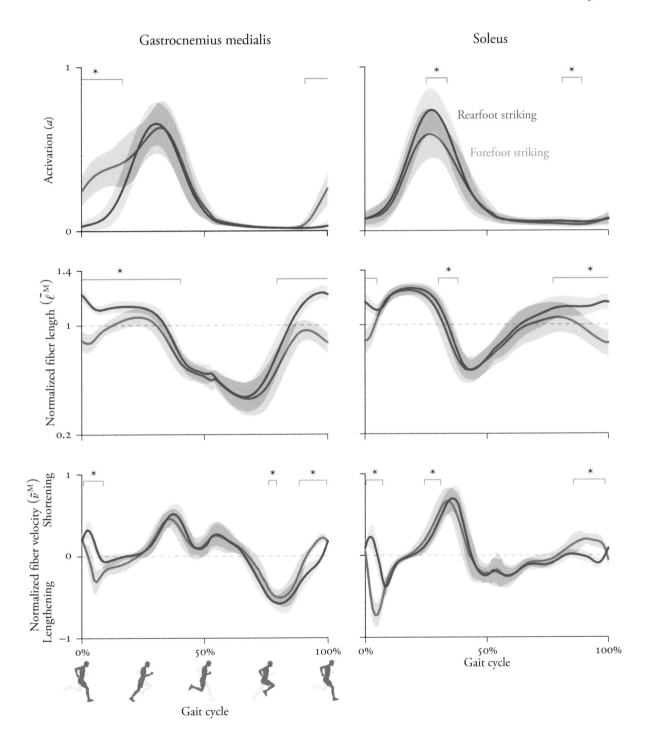

stance, the plantarflexor fibers shorten when rearfoot striking but lengthen when forefoot striking.

Interestingly, despite the springy appearance of forefoot striking, this foot-strike pattern does not affect the amount of energy stored in the gastrocnemius tendon (the apparent difference is not statistically significant), and the soleus tendon actually stores *less* energy when forefoot striking (Figure 12.11, top). The total energy stored in the Achilles tendon does not differ between forefoot and rearfoot striking; however, the timing of energy storage and return is strikingly different between foot-strike patterns (Figure 12.11, bottom).

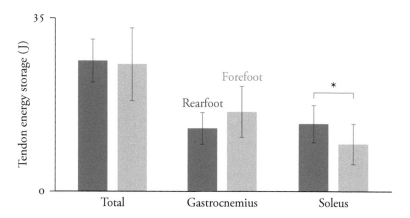

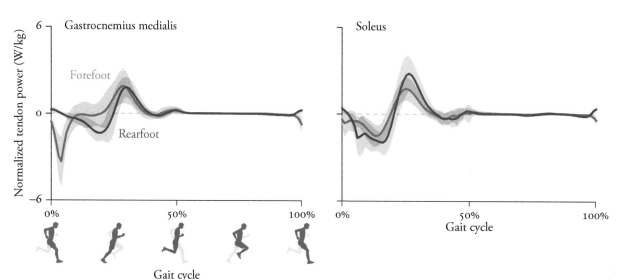

FIGURE 12.11
Tendon elastic energy (top) and power (bottom) for plantarflexor muscles in simulations of running while rearfoot and forefoot striking (n = 16, mean ±1 standard deviation; *p < 0.05). Data from Yong et al. (2020).

Forefoot striking increases the rate at which energy is absorbed by the gastrocnemius fibers in early stance, when the muscle is generating large forces and lengthening. Muscles tend to become injured when they generate large forces while lengthening, and this may explain why some rearfoot-striking runners report calf muscle soreness when attempting to convert to forefoot striking. My recommendation to runners who are planning to convert to forefoot striking is to adopt a progressive eccentric gastrocnemius strengthening program prior to converting, to avoid muscle injury.

Device-assisted running

Assistive devices can interact with our musculoskeletal system to restore mobility and enhance performance. Among the many potential applications of this technology is reducing the metabolic cost of running. Thomas Uchida used muscle-driven simulations of running to explore the biomechanical and energetic effects of assistive devices (Uchida et al., 2016b). The simulated devices were massless and could deliver large torques instantaneously, with precise timing, and directly to the skeleton. The simulations enabled comparison of assistive devices without building prototypes, and in isolation from confounding factors such as imperfect transmission of forces to the body. Uchida used muscle-driven simulations of 10 athletes running on a treadmill at 2 and 5 m/s. He augmented the musculoskeletal model shown in Figure 12.1 with ideal assistive devices that provide flexion and extension moments at the hips, knees, and ankles, and then determined the patterns of assistance that minimize the sum of squared simulated muscle activations. The metabolic cost of running with and without assistance was estimated using a model of the energy consumed by muscle fibers as a function of their excitation, activation, length, and velocity (Umberger et al., 2003; Uchida et al., 2016a).

The simulations predicted a substantial reduction in metabolic cost in all assistance scenarios at both running speeds (Figure 12.12). When running at 2 m/s, the ankle, knee, and hip devices were equally effective, reducing average metabolic power by 1.5 W/kg, about one-quarter of the metabolic power consumed during unassisted running. When running at 5 m/s, however, the hip device reduced metabolic cost significantly more than the ankle or knee devices (p < 0.002). This result suggests that assistive devices should adapt to the user's running speed by redistributing device power from the ankle to the hip as speed increases.

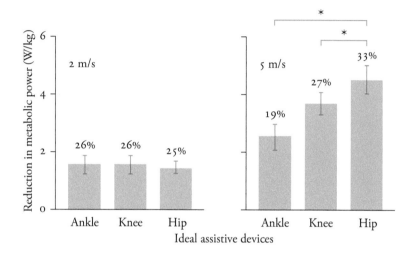

FIGURE 12.12
Reductions in the energy expended by lower-extremity muscles when a musculoskeletal model was assisted by ideal flexion–extension assistive devices (the shaded column represents the mean; the line represents standard deviation; n = 10; *p < 0.002). Average power was normalized by subject mass (W/kg), and percentages were computed relative to the cost of unassisted running at each speed. Adapted from Uchida et al. (2016b).

Figure 12.13 shows the effect of each assistive device on the coordination of nine muscles when running at 5 m/s. Muscle activations generally decreased in muscles crossing assisted joints; however, activations also decreased in muscles crossing *unassisted* joints, and some activations *increased* when assistance was added. With assistance added at the ankle, the soleus and tibialis anterior activations decreased dramatically throughout the gait cycle because these muscles generate only ankle moments. The activation of the gastrocnemius decreased only partially during stance; although it was no longer responsible for generating an ankle plantarflexion moment, it was still recruited to generate a flexion moment at the (unassisted) knee. What remained of the original gastrocnemius activation roughly reflects its activity attributable to generating a knee flexion moment during stance. With assistance added at the knee, we observed an increase in the rectus femoris activation during early swing to take advantage of its relatively high

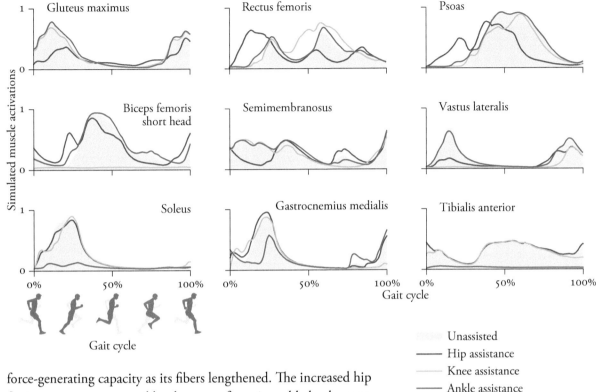

FIGURE 12.13
Activations of nine representative lower-extremity muscles in simulations of running at 5 m/s with ideal flexion–extension assistive devices (mean; n = 10). Adapted from Uchida et al. (2016b).

force-generating capacity as its fibers lengthened. The increased hip flexion moment generated by the rectus femoris enabled a decrease in the activity of the psoas. Complex changes in muscle activity were observed when assistance was added at the hip.

Compensations to assistance can be complex. Muscle-driven simulations like these are an important complement to experiments because they provide biomechanical explanations for observed effects and suggest hypotheses for future experimental studies. Our simulations demonstrated the importance of considering muscle coordination throughout the lower limbs when designing assistance strategies. As we saw for device-assisted walking in Chapter 11, the optimal device torque profiles generally differ from the corresponding net joint moments generated by muscles during unassisted running. This finding was confirmed by a team led by Connor Walsh at Harvard, who used our results to determine what hip extension torques their assistive device should apply to reduce the metabolic cost of running (Figure 12.14). When applying

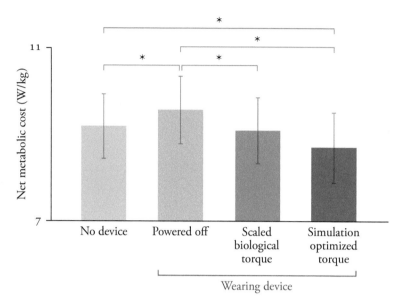

FIGURE 12.14
The hip extension device of Lee et al. was more effective at reducing the metabolic cost of running when it applied a torque that was informed by muscle-driven simulations rather than net joint moments. Metabolic power consumption is shown relative to the power consumed during standing, normalized by body mass, and averaged over 9 subjects during running at 2.5 m/s (mean ±1 standard deviation; *$p \leq 0.05$). Adapted from Lee et al. (2017).

a scaled version of the net hip extension moment generated by muscles when unassisted, their device reduced net metabolic power consumption by only 0.11 W/kg. A torque profile informed by the results of our simulations was four times more effective.

Springs to enhance running

Elliot Hawkes, an avid runner and former student in my Biomechanics of Movement course, was riding his bike in Golden Gate Park one afternoon. The park has concentric tracks, with an inner track for bikes and an outer track for runners. Riding in circles was boring, so Elliot naturally turned his attention to the runners. From the reference frame of his bike, the runners appeared almost stationary—except for their legs, which looked like pendula oscillating back and forth under a fixed trunk. Elliot thought he might be able to save the runners energy by joining their legs with a spring. The next day, Elliot and his lab mate Cole Simpson, who had been thinking along similar lines, tied some surgical tubing to their shoes and went for a brief run. It felt amazing. This quick investigation inspired a two-year team project to explore how we might decrease the energetic cost of running with this simple device.

The team made springs from latex tubing and attached each end to the top of runners' shoes (Figure 12.15). The spring must be long enough that it does not apply elastic forces when the feet pass each other and does not break when the feet are far apart, yet short enough that it does not become entangled. The stiffness of the initial device was 125 N/m, with a free length equal to 25 percent of the participant's leg length. The mass was about 0.02 kg. We tested many springs, but none offered a substantial advantage over this initial design. The student inventors each ran a 6 km loop around the Stanford campus without tripping (which was an initial concern).

FIGURE 12.15
Time-lapse photographs of a runner using a spring connecting the legs to improve running efficiency. The spring generates force when the legs are separated. Photo courtesy of Cara Welker.

We measured metabolic energy consumption while 12 healthy participants ran at 2.7 m/s on a treadmill with and without shoes connected by springs. Over two testing days, the participants each completed four 10-minute runs with the spring, interspersed with four 10-minute natural runs without the spring as experimental controls. During the first trial, participants showed no metabolic savings when running with the spring compared to natural running. However, by the end of the second trial, participants expended 3.8 ± 5.4 percent less energy ($p = 0.034$) during spring-assisted running compared to natural running. Metabolic savings increased with practice, and all participants ran using less energy by the end of the second day of testing. The average energy savings was 6.4 ± 2.8 percent (Simpson et al., 2019).

To explore the mechanism by which the spring reduced energy, we developed an OpenSim model of running with and without a spring connecting the legs. The model revealed that the spring produced moments about the hip and knee, which reduced the hip and knee moments generated by muscles (Figure 12.16). The spring improves running economy through a complex mechanism that produces savings beyond those associated with leg swing. When runners learn to use the spring, they adopt a running gait with shorter strides and higher stride frequency. The shorter and faster strides, in turn, reduce the work required to redirect the center of

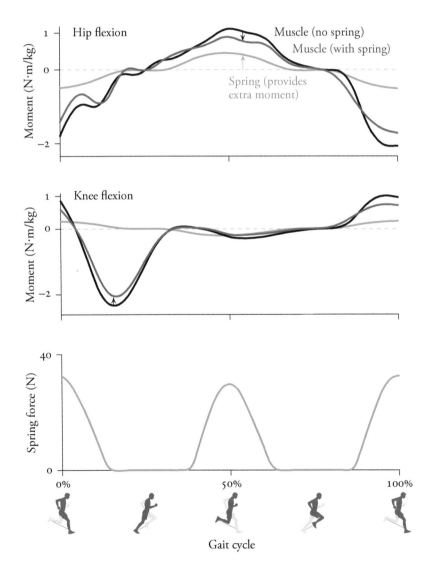

FIGURE 12.16
Mechanism of energetic savings when running with a spring connecting the legs. Moments generated by muscles during swing were reduced due to the assistance of the spring. Muscle-generated moments during stance were also reduced, possibly due to the increased stride frequency. Adapted from Simpson et al. (2019).

mass during stance. Therefore, the spring reduces the work not only during swing but also during stance.

A spring connecting the legs could serve as an inexpensive assistive device to improve human running performance, or a simple intervention to further explore human–machine interactions. Elliot Hawkes' experience of conceiving of this invention from his bicycle reminds us that a new perspective can lead to creative insights.

13 Moving Forward

We must walk consciously only part way toward our goal, and then leap in the dark to our success.
—Henry David Thoreau

THE FIELD OF MOVEMENT BIOMECHANICS has made great strides, but there is more ground to cover. In this book, we have seen how measurements, models, and computational tools can help us understand human movement. We have shown how biomechanical models have been used to design robots, running tracks, and exoskeletons. Biomechanics has enabled us to design factories that protect workers from injuries and to create joint replacements that enable people with diseased joints to walk without pain.

These triumphs have improved the lives of millions of people, but we must also be aware of the limits of current knowledge and technology. New developments in technology will enable us to make further positive impacts in people's lives; the increasing functionality of prosthetic limbs is a good example. Limitations of knowledge are harder to overcome. More than anything else, overcoming these limitations requires our creativity and, as Thoreau suggests, the courage to take a leap in the dark.

I would like to start this chapter with an example of a creative leap that one of my collaborators took, to great success. I will move on to paint a picture of the future I envision for the field of biomechanics—a picture that is surely incomplete. Your vision will add color and texture to the canvas of possibilities, where there is infinite space for you to create your masterpiece.

Wearable technology

Kate Rosenbluth was working in my laboratory when she met a man at the Stanford Hospital who was unable to write a note to his wife or drink coffee with a friend because his hands shook from a condition called essential tremor. This is an enormously frustrating condition that is nevertheless very common. People who develop essential tremor in their fifties and sixties find themselves unable to carry out activities of daily living because their hands shake.

The man with essential tremor told Kate that the drugs he had tried did not reduce his hand tremor and had terrible side effects. His only remaining option was a surgery to place stimulating electrodes in his brain. This therapy, called deep brain stimulation, has proved very effective for reducing tremor, but it requires brain surgery. With brain surgery comes a risk of serious complications. It is not a decision to take lightly.

Kate knew that his shaking hands were likely caused by oscillatory neural activity within specific areas of the brain, including a region called the ventral intermediate nucleus of the thalamus. Kate found articles showing that electrical stimulation of the median nerve at the wrist evokes neural activity within the ventral intermediate nucleus. She reasoned that stimulation of the sensory nerves at the wrist might produce activity in this brain region and thus reduce hand tremor without surgery. As an experiment, she decided to try stimulating the median nerve in a few willing participants with essential tremor.

It was a leap in the dark. We don't know exactly what causes the oscillatory neural activity in the brain, and we didn't know whether electrical stimulation at the wrist would affect it. But to our (and our participants') delight, we saw dramatic reductions in their hand tremor after stimulating their median nerves. These preliminary data encouraged us to push forward. We continued experimenting and learning about this new technology. For example, we saw the best results when stimulating the nerves at the same frequency as the tremor. Kate teamed up with Serena Wong and others to build a wearable motion sensor and neurostimulator in the form of a

Wearable neurostimulator

Before stimulation

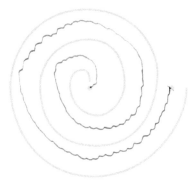
After stimulation

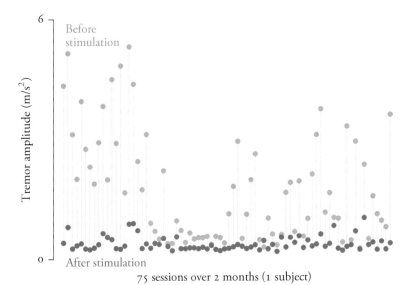

FIGURE 13.1
Wearable motion sensor and neurostimulator for reducing hand tremor. A wearable stimulator from Cala Health (top left) reduces hand tremor substantially (bottom). A patient's spiral drawing improves dramatically after 40 minutes of neurostimulation therapy (top right). Cala Health, Inc. is a Stanford University spin-out company cofounded by Kate Rosenbluth, Serena Wong, and Scott Delp.

wristwatch that records the frequency of a patient's hand tremor and stimulates the nerves in the wrist at their tremor frequency. It was incredibly rewarding to find that many people who use this device experience a meaningful reduction in hand tremor (Lin et al., 2018). They recover the ability to write a note, drink coffee, and perform other activities that were impossible with shaking hands (Figure 13.1).

This new treatment reduces hand tremor without medication and without surgery. It was made possible by Kate's creativity and environmental, social, and personal factors that motivate healthy behaviors and identify new ways to improve health.

Mobile apps, wearables, and the massive amounts of data they are collecting have transformational potential. However, effective analysis of these data requires expertise in biomechanics, data science, and public health. Few researchers are cross-trained in these areas, making collaboration and communication between disciplines difficult. I have tried to gain experience in data science and public health so that I can engage with researchers in these domains and mentor students who want to cross these disciplinary boundaries. This is an area in which you can have great impact, and I encourage you to dive in.

Large-scale studies require data and sophisticated tools to gain insights from these data. Unfortunately, most research labs do not share their data or tools. Companies maintain databases that characterize the activities of millions of individuals, but these databases are rarely shared with researchers for analysis. The future of biomechanics—and, indeed, the health of millions of individuals—calls for us to liberate these data and develop methods to gain insights from these rich datasets.

Modern statistics and machine learning

The conclusions of nearly all biomedical research studies are based on hypothesis testing using parametric tests, such as the Student's *t*-test. The current surge of data, however, presents new challenges for many biomedical disciplines, including biomechanics. Data characterizing human movement are high-dimensional, heterogeneous, and growing in volume with wearable sensing. Traditional statistical methods limit the insights we can gain from such data. New statistical methods will transform biomechanics, as they have done with autonomous driving, image recognition, and automated cancer detection.

One popular method to unlock the power of large datasets is machine learning. Eni Halilaj searched for research articles that used machine learning to study movement biomechanics

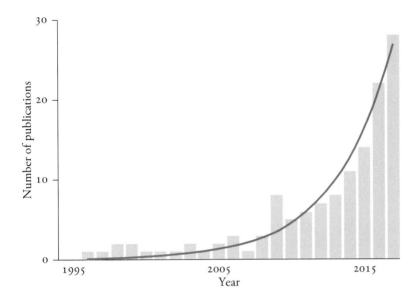

FIGURE 13.3
Use of data science methods in human movement biomechanics studies has increased dramatically in recent years. Adapted from Halilaj et al. (2018).

and discovered a rapid rise in the number of publications in this area (Figure 13.3). I expect this trend to continue as more people engage in collaboration and cross-training between biomechanics and machine learning.

A study by Apoorva Rajagopal and colleagues illustrates the impact of combining modern statistical methods and biomechanical modeling. Apoorva's study examined surgical correction of equinus gait (walking on the toes), one of the most common gait abnormalities in children with cerebral palsy. She wanted to know whether we could use estimates of gastrocnemius length during gait to identify limbs whose ankle kinematics were likely to improve following gastrocnemius lengthening surgery (Figure 13.4). She analyzed gait data from 891 limbs that underwent surgery and categorized outcomes based on the presurgical and postsurgical ankle kinematics. Limbs with a short gastrocnemius that received a gastrocnemius lengthening surgery ("case" limbs) were far more likely to have a good surgical outcome than limbs whose gastrocnemius was not short yet still received a lengthening surgery ("overtreated" limbs). The difference was staggering: 71 percent of case limbs had good outcomes at the follow-up gait visit compared

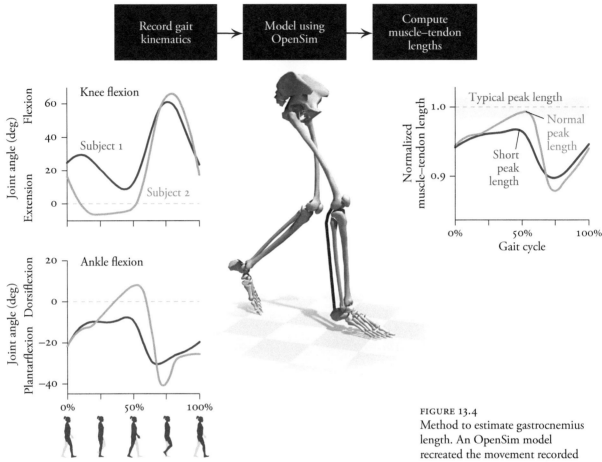

FIGURE 13.4
Method to estimate gastrocnemius length. An OpenSim model recreated the movement recorded with motion capture (left) and then computed muscle–tendon length of the gastrocnemius medialis (right). These lengths were normalized to the average peak length computed for typical gait. Limbs whose peak length was at least 2 standard deviations below the typical mean (i.e., below the shaded band in the right panel) were categorized as having "short" gastrocnemius muscles. Subject 1 (blue lines) had a short peak gastrocnemius length and Subject 2 (orange lines) did not. Adapted from Rajagopal et al. (2019).

to only 33 percent of overtreated limbs (Figure 13.5). Her results suggest that estimates of gastrocnemius length should be used to identify good candidates for surgery.

Modeling neuromuscular control to predict movement

Modeling how the brain coordinates movement is a grand challenge in biomechanics. Computational modeling has mostly been limited to reproducing observed movements to study quantities that cannot be measured, such as muscle forces and joint loads.

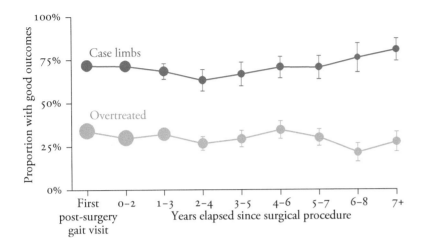

FIGURE 13.5
Long-term outcomes following gastrocnemius lengthening surgery. Case limbs underwent surgery consistent with the musculoskeletal model and had better outcomes than overtreated limbs. The observations were binned based on years elapsed following the surgery, and the rate of good outcome in each bin was computed. Bins overlap in time. The error bars indicate 1 standard deviation; areas of the circles are proportional to the number of observations. Adapted from Rajagopal et al. (2019).

Modern computational techniques have enabled development of predictive simulations, which produce simulations of movement without first collecting experimental data. Predictive simulations use optimization to achieve high-level objectives, such as minimizing a model's energy expenditure during gait, while adhering to physical constraints like Newton's laws and physiological constraints like muscle activation delays. The variables in the optimization problem describe how to coordinate the model's muscles. The solution produces a simulation that represents movement observed in nature as long as the optimization problem captures the salient qualities of real movement. Predictive simulations will allow us to design better prostheses and surgeries that maximize an individual's function.

To generate predictive simulations, we must solve high-dimensional optimization problems. One strategy for solving these problems is to use reinforcement learning, which can be applied without knowledge of an underlying model and without proposing the structure of a controller: the optimizer simply learns how to generate model inputs (in this case, muscle excitations over time) that produce desirable outputs. Łukasz Kidziński (2018) ran a fantastic competition in conjunction with the past two Neural Information Processing Systems conferences in which over a thousand participants used reinforcement learning to solve classical problems in motor control. For example, competitors were challenged to develop controllers that would enable a musculoskeletal model to run over uneven terrain. The results were amazing! The competitors proved that reinforcement learning techniques can synthesize natural movements like walking and running. Their optimizers also discovered a variety of bizarre hopping and skipping gaits. This project was successful because of the collaboration between biomechanists and computer scientists, and the availability of powerful open-source software like OpenSim and the reinforcement learning environment from CrowdAI.

Direct collocation is another powerful method to solve high-dimensional optimization problems in biomechanics. Direct collocation converts the equations of motion into algebraic constraint equations, effectively computing all time frames simultaneously. Solving optimization problems is challenging, and researchers frequently wait for days, or even weeks, for the results of an optimization. Direct collocation has been used to solve optimal control problems in biomechanics much faster than traditional methods and will enable broader use of predictive simulations. I am certain that direct collocation will have a great impact on the future of biomechanics, and I encourage you to learn more about it.

Motivating movement

We all know that physical activity is good for us. As Hippocrates famously said, "Walking is man's best medicine." Yet nearly half of adults fail to get enough activity to maintain their health, and this inactivity is costing us dearly. Worldwide, five million deaths each year are due to conditions like heart disease, stroke, and diabetes, and many of these could be prevented just by increasing physical activity (Lee et al., 2012). Technology that motivates healthy behaviors—especially well-established, low-cost behaviors like exercise—could improve the health of millions of people and lower the economic burden of physical inactivity.

Smartphones and wearable sensors have the potential to fulfill this vision and revolutionize healthcare. I have already mentioned that these devices provide vast amounts of data about physical activity, and smartphones enable near-continuous interaction with users for motivating healthy behaviors. However, smartphones and other devices have not improved health—and some have had the opposite effect (Jakicic et al., 2016). Clearly, technological capability alone is insufficient to solve this problem. If we want to make these devices into active tools for improving public health, not just for gathering information, we have to understand how people interact with technology. Motivating individuals is more complex than merely developing a nifty new gadget.

Physical activity is a powerful, low-cost medicine, but we need precise and engaging tools to motivate it. Ideas from diverse disciplines such as behavioral psychology, human–computer interaction theory, and biomechanics have been successful on a small scale; however, these innovations have rarely been incorporated into mobile health apps. Biomechanists, in collaboration with experts in other disciplines, must look for new ways to motivate physical activity that will reach millions of people around the world and address areas of great need, such as neurological disability. Our vision is to achieve effective yet affordable healthcare that is deployable on a global scale.

Open science

Open science is the best way to advance our field. As a community, we must make publications, data, models, and software freely available to everyone. Immense amounts of high-fidelity movement data have been collected from healthy individuals and those with every imaginable movement impairment. But the data are hidden on computer hard drives and unavailable to the community. Anonymizing and releasing these data would enable biomechanical studies of unprecedented scale. To encourage data sharing and open science, my colleagues and I have created a website for sharing biomechanical data, models, and software—simtk.org—which hosts over 1,000 projects with shared resources. The models and data presented in this book are available on this website. The biomechanics culture must continue to change to make data sharing the norm.

FIGURE 13.6
OpenSim model of an ostrich, the fastest biped on the planet. Model from Rankin et al. (2016).

Sharing software can also accelerate research. Biomechanical simulation is rapidly growing in popularity, in part because it is a powerful complement to experimental approaches, but also because established software tools are now available. As we have seen throughout the book, the OpenSim software enables users to build biomechanical models, simulate musculoskeletal dynamics, predict novel movements, and disseminate new computational tools. Jeff Rankin, Jonas Rubenson, and John Hutchinson (2016) used OpenSim to study the mechanisms responsible for the impressive speed, agility, and efficiency of the ostrich, the fastest-running biped (Figure 13.6). Taymaz Homayouni and colleagues (2015) used OpenSim to prototype new implantable mechanisms to restore function in patients with partial paralysis of the arm. Matt DeMers, Jennifer Hicks, and I used OpenSim to compare the effectiveness of reflexes and preparatory muscle coactivation in preventing ankle injuries (Figure 13.7), a study that would be

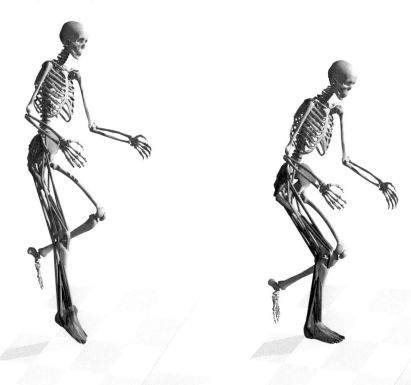

FIGURE 13.7
Frames from a simulation of landing on an incline to study ankle injuries. OpenSim model from DeMers et al. (2017).

too dangerous to perform experimentally. These are just a few of the hundreds of studies that have been enabled by the OpenSim software, and I am delighted to see the OpenSim community continuing to grow.

In addition to open-source software, open access to computational models of musculoskeletal structures has enabled a wide range of studies. For example, Sam Hamner's model to simulate running (Figure 12.4), Miguel Christophy's model of the lumbar spine (Christophy et al., 2012), and Kate Saul's model of the upper extremity (Figure 10.6) have each been used by thousands of researchers and students. Continuing the trend toward open science will enable investigation of increasingly complex problems that can only be solved through the combined effort of individuals with diverse expertise. The OpenSim project provides a collaborative research platform that supports a diverse, global, and expanding community that is working to solve the most important problems in biomechanics (Figure 13.8). I hope you will join this team.

FIGURE 13.8
Locations of visitors to the OpenSim documentation in a recent one-year period. Since its launch in 2012, the OpenSim documentation wiki has been visited by over 25,000 users from around the world per year. Adapted from Seth, Hicks, Uchida et al. (2018).

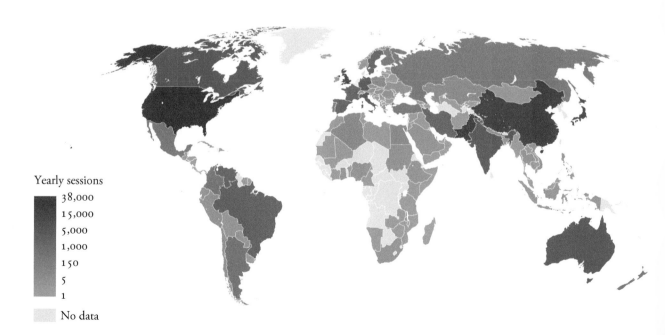

Taking the baton

I hope this book has increased your appreciation of the elegance and complexity of human movement. Even the limited material we have covered is the culmination of research that has taken thousands of individuals hundreds of years to develop. It's an amazing team effort. As Walt Whitman wrote, "the powerful play goes on, and you may contribute a verse." Indeed, the principles you have learned by working through this book set the stage for you to participate in the field. You can read news articles and journal papers about running shoes and robots with a new perspective. You can attend a seminar or conference, or just talk to your friend about the way we move. With a little creativity and a bit of hard work, you can make meaningful contributions to the world through biomechanics. Let's keep moving!

Symbols

Conventions

$\underline{x}$	vector
X	matrix
X, Y, Z	axes of a reference frame
$\hat{x}, \hat{y}, \hat{z}$	unit vectors along X, Y, and Z
$\dot{x}(t)$ or $\dot{x}$	$\frac{d}{dt}x(t)$
$\ddot{x}(t)$ or $\ddot{x}$	$\frac{d^2}{dt^2}x(t)$
p_A	point p expressed in frame A
$^A R_B$	3×3 rotation matrix that expresses p_B in frame A: $p_A = {}^A R_B\, p_B$
$^A T_B$	4×4 transformation matrix that expresses p_B in frame A

Acronyms

ACL	anterior cruciate ligament
ADP / ATP	adenosine diphosphate / adenosine triphosphate
CMA-ES	covariance matrix adaptation evolution strategy
COM	center of mass
COP	center of pressure
EMG	electromyography
GRF	ground reaction force(s)
IMU	inertial measurement unit
MRI	magnetic resonance imaging
PCSA	physiological cross-sectional area

Nomenclature

θ	angle
σ^M	muscle stress
σ_o^M	muscle specific tension (peak isometric stress)
τ_A	muscle activation time constant
τ_D	muscle deactivation time constant
ϕ	muscle fiber pennation angle
ϕ_o	muscle fiber pennation angle when $\ell^M = \ell_o^M$
a	acceleration; muscle activation ($0 \leq a \leq 1$)
E_{kf}	kinetic energy in the forward direction
E_{pg}	gravitational potential energy
F	force
F^{GRF}	ground reaction force
$F^M \{\tilde{F}^M\}$	muscle force {muscle force normalized by F_o^M}
F_o^M	maximum isometric muscle force
$F^T \{\tilde{F}^T\}$	tendon force {tendon force normalized by F_o^M}
$f^L(\tilde{\ell}^M)$	normalized active force–length relation of the muscle
$f^{PE}(\tilde{\ell}^M)$	normalized passive force–length relation of the muscle
$f^T(\tilde{\ell}^T)$	normalized force–length relation of the tendon
$f^V(\tilde{v}^M)$	normalized force–velocity relation of the muscle
Fr	Froude number ($0 \leq \text{Fr} \leq 1$)
g	gravitational acceleration (≈ 9.81 m/s^2)
h	height of pennated muscle model
I	inertia
J	objective function; Jacobian
ℓ	length
$\ell^M \{\tilde{\ell}^M\}$	muscle length {muscle length normalized by ℓ_o^M}
ℓ_o^M	length at which muscle develops maximum isometric active force ("optimal" muscle length)
ℓ^{MT}	length of muscle–tendon actuator
ℓ^S	sarcomere length
ℓ_o^S	length at which sarcomere develops maximum isometric active force ("optimal" sarcomere length)
$\ell^T \{\tilde{\ell}^T\}$	tendon length {tendon length normalized by ℓ_s^T}

ℓ_s^T	shortest length at which tendon generates a tensile force (tendon "slack" length)
m	mass
M	moment; mass matrix
q	generalized coordinate
r	moment arm
$\underline{r}$	position vector
t	time
T	torque
u	muscle excitation ($0 \leq u \leq 1$)
v	velocity
$v^M \{\tilde{v}^M\}$	muscle shortening velocity {the same normalized by v_{max}^M}, where $v^M(t) = -\dot{\ell}^M(t)$
v_{max}^M	maximum muscle contraction velocity
v^{MT}	lengthening velocity of muscle–tendon actuator
v_{max}^S	maximum sarcomere contraction velocity
v^T	tendon lengthening velocity

References

Abbott BC, Bigland B, Ritchie JM. The physiological cost of negative work. *The Journal of Physiology* 117: 380–390, 1952.

ACCAD. *Open Motion Project.* Advanced Computing Center for the Arts and Design, The Ohio State University, 2018.

Alexander RM. Walking and running: legs and leg movements are subtly adapted to minimize the energy costs of locomotion. *American Scientist* 72: 348–354, 1984.

Althoff T, Sosič R, Hicks JL, King AC, Delp SL, Leskovec J. Large-scale physical activity data reveal worldwide activity inequality. *Nature* 547: 336–339, 2017.

An KN, Takahashi K, Harrigan TP, Chao EY. Determination of muscle orientations and moment arms. *Journal of Biomechanical Engineering* 106: 280–282, 1984.

Anderson FC, Arnold AS, Pandy MG, Goldberg SR, Delp SL. Simulation of walking. In J Rose, JG Gamble (Eds.), *Human Walking* (3rd ed., pp. 195–210). Philadelphia, PA: Lippincott Williams & Wilkins, 2006.

Arnold AS, Liu MQ, Schwartz MH, Õunpuu S, Delp SL. The role of estimating muscle-tendon lengths and velocities of the hamstrings in the evaluation and treatment of crouch gait. *Gait & Posture* 23: 273–281, 2006.

Arnold AS, Thelen DG, Schwartz MH, Anderson FC, Delp SL. Muscular coordination of knee motion during the terminal-swing phase of normal gait. *Journal of Biomechanics* 40: 3314–3324, 2007.

Arnold EM, Hamner SR, Seth A, Millard M, Delp SL. How muscle fiber lengths and velocities affect muscle force generation as humans walk and run at different speeds. *The Journal of Experimental Biology* 216: 2150–2160, 2013.

Ashby BM, Heegaard JH. Role of arm motion in the standing long jump. *Journal of Biomechanics* 35: 1631–1637, 2002.

Bartlett JL, Kram R. Changing the demand on specific muscle groups affects the walk–run transition speed. *The Journal of Experimental Biology* 211: 1281–1288, 2008.

Bates NA, Hewett TE. Motion analysis and the anterior cruciate ligament: classification of injury risk. *The Journal of Knee Surgery* 29: 117–125, 2016.

Biswas A, Oh PI, Faulkner GE, Bajaj RR, Silver MA, Mitchell MS, Alter DA. Sedentary time and its association with risk for disease incidence, mortality, and hospitalization in adults: a systematic review and meta-analysis. *Annals of Internal Medicine* 162: 123–132, 2015.

Blemker SS, Delp SL. Three-dimensional representation of complex muscle architectures and geometries. *Annals of Biomedical Engineering* 33: 661–673, 2005.

Browning RC, Baker EA, Herron JA, Kram R. Effects of obesity and sex on the energetic cost and preferred speed of walking. *Journal of Applied Physiology* 100: 390–398, 2006.

Buchanan TS, Delp SL, Solbeck JA. Muscular resistance to varus and valgus loads at the elbow. *Journal of Biomechanical Engineering* 120: 634–639, 1998.

Cavagna GA, Thys H, Zamboni A. The sources of external work in level walking and running. *The Journal of Physiology* 262: 639–657, 1976.

Cazzola D, Holsgrove TP, Preatoni E, Gill HS, Trewartha G. Cervical spine injuries: a whole-body musculoskeletal model for the analysis of spinal loading. *PLoS ONE* 12: e0169329, 2017.

Christophy M, Faruk Senan NA, Lotz JC, O'Reilly OM. A musculoskeletal model for the lumbar spine. *Biomechanics and Modeling in Mechanobiology* 11: 19–34, 2012.

Collins SH, Adamczyk PG, Kuo AD. Dynamic arm swinging in human walking. *Proceedings of the Royal Society B* 276: 3679–3688, 2009.

Collins SH, Ruina A, Tedrake R, Wisse M. Efficient bipedal robots based on passive-dynamic walkers. *Science* 307: 1082–1085, 2005.

Collins SH, Wiggin MB, Sawicki GS. Reducing the energy cost of human walking using an unpowered exoskeleton. *Nature* 522: 212–215, 2015.

Collins SH, Wisse M, Ruina A. A three-dimensional passive-dynamic walking robot with two legs and knees. *The International Journal of Robotics Research* 20: 607–615, 2001.

Dawson TJ. Kangaroos. *Scientific American* 237: 78–89, 1977.

Delp SL, Anderson FC, Arnold AS, Loan P, Habib A, John CT, Guendelman E, Thelen DG. OpenSim: open-source software to create and analyze dynamic simulations of movement. *IEEE Transactions on Biomedical Engineering* 54: 1940–1950, 2007.

Delp SL, Loan JP. A graphics-based software system to develop and analyze models of musculoskeletal structures. *Computers in Biology and Medicine* 25: 21–34, 1995.

Delp SL, Loan JP, Hoy MG, Zajac FE, Topp EL, Rosen JM. An interactive graphics-based model of the lower extremity to study orthopaedic surgical procedures. *IEEE Transactions on Biomedical Engineering* 37: 757–767, 1990.

Dembia CL, Silder A, Uchida TK, Hicks JL, Delp SL. Simulating ideal assistive devices to reduce the metabolic cost of walking with heavy loads. *PLoS ONE* 12: e0180320, 2017.

DeMers MS, Hicks JL, Delp SL. Preparatory co-activation of the ankle muscles may prevent ankle inversion injuries. *Journal of Biomechanics* 52: 17–23, 2017.

Domalain M, Bertin A, Daver G. Was *Australopithecus afarensis* able to make the Lomekwian stone tools? Towards a realistic biomechanical simulation of hand force capability in fossil hominins and new insights on the role of the fifth digit. *Comptes Rendus Palevol* 16: 572–584, 2017.

Donelan JM, Kram R, Kuo AD. Mechanical work for step-to-step transitions is a major determinant of the metabolic cost of human walking. *The Journal of Experimental Biology* 205: 3717–3727, 2002.

Donnelly CJ, Lloyd DG, Elliott BC, Reinbolt JA. Optimizing whole-body kinematics to minimize valgus knee loading during sidestepping: implications for ACL injury risk. *Journal of Biomechanics* 45: 1491–1497, 2012.

Dorn TW, Schache AG, Pandy MG. Muscular strategy shift in human running: dependence of running speed on hip and ankle muscle performance. *The Journal of Experimental Biology* 215: 1944–1956, 2012.

Dorn TW, Wang JM, Hicks JL, Delp SL. Predictive simulation generates human adaptations during loaded and inclined walking. *PLoS ONE* 10: e0121407, 2015.

Einstein A. On the method of theoretical physics. *Philosophy of Science* 1: 163–169, 1934.

Engsberg JR, Ross SA, Collins DR. Increasing ankle strength to improve gait and function in children with cerebral palsy: a pilot study. *Pediatric Physical Therapy* 18: 266–275, 2006.

Farley CT, Glasheen J, McMahon TA. Running springs: speed and animal size. *The Journal of Experimental Biology* 185: 71–86, 1993.

Farris DJ, Sawicki GS. Human medial gastrocnemius force–velocity behavior shifts with locomotion speed and gait. *Proceedings of the National Academy of Sciences of the USA* 109: 977–982, 2012.

Fox MD, Reinbolt JA, Õunpuu S, Delp SL. Mechanisms of improved knee flexion after rectus femoris transfer surgery. *Journal of Biomechanics* 42: 614–619, 2009.

Fregly BJ, Reinbolt JA, Rooney KL, Mitchell KH, Chmielewski TL. Design of patient-specific gait modifications for knee osteoarthritis rehabilitation. *IEEE Transactions on Biomedical Engineering* 54: 1687–1695, 2007.

Geyer H, Seyfarth A, Blickhan R. Compliant leg behaviour explains basic dynamics of walking and running. *Proceedings of the Royal Society B* 273: 2861–2867, 2006.

Goldberg SR, Anderson FC, Pandy MG, Delp SL. Muscles that influence knee flexion velocity in double support: implications for stiff-knee gait. *Journal of Biomechanics* 37: 1189–1196, 2004.

Gordon AM, Huxley AF, Julian FJ. The variation in isometric tension with sarcomere length in vertebrate muscle fibres. *The Journal of Physiology* 184: 170–192, 1966.

Gregory CM, Bickel CS. Recruitment patterns in human skeletal muscle during electrical stimulation. *Physical Therapy* 85: 358–364, 2005.

Halilaj E, Rajagopal A, Fiterau M, Hicks JL, Hastie TJ, Delp SL. Machine learning in human movement biomechanics: best practices, common pitfalls, and new opportunities. *Journal of Biomechanics* 81: 1–11, 2018.

Hamner SR, Delp SL. Muscle contributions to fore-aft and vertical body mass center accelerations over a range of running speeds. *Journal of Biomechanics* 46: 780–787, 2013.

Hamner SR, Seth A, Delp SL. Muscle contributions to propulsion and support during running. *Journal of Biomechanics* 43: 2709–2716, 2010.

Handsfield GG, Meyer CH, Hart JM, Abel MF, Blemker SS. Relationships of 35 lower limb muscles to height and body mass quantified using MRI. *Journal of Biomechanics* 47: 631–638, 2014.

Henneman E, Somjen G, Carpenter DO. Functional significance of cell size in spinal motoneurons. *Journal of Neurophysiology* 28: 560–580, 1965.

Hernandez F, Wu LC, Yip MC, Laksari K, Hoffman AR, Lopez JR, Grant GA, Kleiven S, Camarillo DB. Six degree-of-freedom measurements of human mild traumatic brain injury. *Annals of Biomedical Engineering* 43: 1918–1934, 2015.

Hicks JL, Uchida TK, Seth A, Rajagopal A, Delp SL. Is my model good enough? Best practices for verification and validation of musculoskeletal models and simulations of movement. *Journal of Biomechanical Engineering* 137: 020905, 2015.

Hill AV. The heat of shortening and the dynamic constants of muscle. *Proceedings of the Royal Society B* 126: 136–195, 1938.

Homayouni T, Underwood KN, Beyer KC, Martin ER, Allan CH, Balasubramanian R. Modeling implantable passive mechanisms for modifying the transmission of forces and movements between muscle and tendons. *IEEE Transactions on Biomedical Engineering* 62: 2208–2214, 2015.

Hoogkamer W, Kipp S, Frank JH, Farina EM, Luo G, Kram R. A comparison of the energetic cost of running in marathon racing shoes. *Sports Medicine* 48: 1009–1019, 2018.

Horita T, Kitamura K, Kohno N. Body configuration and joint moment analysis during standing long jump in 6-yr-old children and adult males. *Medicine and Science in Sports and Exercise* 23: 1068–1077, 1991.

Hoyt DF, Taylor CR. Gait and the energetics of locomotion in horses. *Nature* 292: 239–240, 1981.

Hutchinson JR, Famini D, Lair R, Kram R. Are fast-moving elephants really running? *Nature* 422: 493–494, 2003.

Hutchinson JR, Garcia M. *Tyrannosaurus* was not a fast runner. *Nature* 415: 1018–1021, 2002.

Huxley AF. Muscle structure and theories of contraction. *Progress in Biophysics and Biophysical Chemistry* 7: 255–318, 1957.

Huxley AF, Niedergerke R. Structural changes in muscle during contraction: interference microscopy of living muscle fibres. *Nature* 173: 971–973, 1954.

Huxley H, Hanson J. Changes in the cross-striations of muscle during contraction and stretch and their structural interpretation. *Nature* 173: 973–976, 1954.

Inman VT, Ralston HJ, Todd F. *Human Walking*. Baltimore, MD: Williams & Wilkins, 1981.

Jakicic JM, Davis KK, Rogers RJ, King WC, Marcus MD, Helsel D, Rickman AD, Wahed AS, Belle SH. Effect of wearable technology combined with a lifestyle intervention on long-term weight loss: the IDEA randomized clinical trial. *Journal of the American Medical Association* 316: 1161–1171, 2016.

Kidziński Ł, Mohanty SP, Ong CF, Hicks JL, Carroll SF, Levine S, Salathé M, Delp SL. Learning to Run Challenge: synthesizing physiologically accurate motion using deep reinforcement learning. In S Escalera, M Weimer (Eds.), *The NIPS '17 Competition: Building Intelligent Systems* (pp. 101–120). Cham, Switzerland: Springer, 2018.

Kulmala J-P, Kosonen J, Nurminen J, Avela J. Running in highly cushioned shoes increases leg stiffness and amplifies impact loading. *Scientific Reports* 8: 17496, 2018.

Kuo AD. A least-squares estimation approach to improving the precision of inverse dynamics computations. *Journal of Biomechanical Engineering* 120: 148–159, 1998.

Kuo AD. Energetics of actively powered locomotion using the simplest walking model. *Journal of Biomechanical Engineering* 124: 113–120, 2002.

Kuo AD, Donelan JM. Dynamic principles of gait and their clinical implications. *Physical Therapy* 90: 157–174, 2010.

Lai A, Schache AG, Lin Y-C, Pandy MG. Tendon elastic strain energy in the human ankle plantar-flexors and its role with increased running speed. *The Journal of Experimental Biology* 217: 3159–3168, 2014.

Lee G, Kim J, Panizzolo FA, Zhou YM, Baker LM, Galiana I, Malcolm P, Walsh CJ. Reducing the metabolic cost of running with a tethered soft exosuit. *Science Robotics* 2: eaan6708, 2017.

Lee I-M, Shiroma EJ, Lobelo F, Puska P, Blair SN, Katzmarzyk PT. Effect of physical inactivity on major non-communicable diseases worldwide: an analysis of burden of disease and life expectancy. *The Lancet* 380: 219–229, 2012.

Lee SSM, Piazza SJ. Built for speed: musculoskeletal structure and sprinting ability. *The Journal of Experimental Biology* 212: 3700–3707, 2009.

Lieber RL. *Skeletal Muscle Structure, Function, and Plasticity: The Physiological Basis of Rehabilitation* (3rd edition). Baltimore, MD: Lippincott Williams & Wilkins, 2010.

Lieber RL, Yeh Y, Baskin RJ. Sarcomere length determination using laser diffraction. Effect of beam and fiber diameter. *Biophysical Journal* 45: 1007–1016, 1984.

Lin PT, Ross EK, Chidester P, Rosenbluth KH, Hamner SR, Wong SH, Sanger TD, Hallett M, Delp SL. Noninvasive neuromodulation in essential tremor demonstrates relief in a sham-controlled pilot trial. *Movement Disorders* 33: 1182–1183, 2018.

Liu MQ, Anderson FC, Schwartz MH, Delp SL. Muscle contributions to support and progression over a range of walking speeds. *Journal of Biomechanics* 41: 3243–3252, 2008.

Llewellyn ME, Barretto RPJ, Delp SL, Schnitzer MJ. Minimally invasive high-speed imaging of sarcomere contractile dynamics in mice and humans. *Nature* 454: 784–788, 2008.

Llewellyn ME, Thompson KR, Deisseroth K, Delp SL. Orderly recruitment of motor units under optical control in vivo. *Nature Medicine* 16: 1161–1165, 2010.

Lyght M, Nockerts M, Kernozek TW, Ragan R. Effects of foot strike and step frequency on Achilles tendon stress during running. *Journal of Applied Biomechanics* 32: 365–372, 2016.

Malcolm P, Derave W, Galle S, De Clercq D. A simple exoskeleton that assists plantarflexion can reduce the metabolic cost of human walking. *PLoS ONE* 8: e56137, 2013.

Martin PE, Rothstein DE, Larish DD. Effects of age and physical activity status on the speed-aerobic demand relationship of walking. *Journal of Applied Physiology* 73: 200–206, 1992.

McDaniel J, Lombardo LM, Foglyano KM, Marasco PD, Triolo RJ. Setting the pace: insights and advancements gained while preparing for an FES bike race. *Journal of NeuroEngineering and Rehabilitation* 14: 118, 2017.

McGeer T. Passive dynamic walking. *The International Journal of Robotics Research* 9: 62–82, 1990.

McMahon TA. *Muscles, Reflexes, and Locomotion*. Princeton, NJ: Princeton University Press, 1984.

McMahon TA, Greene PR. The influence of track compliance on running. *Journal of Biomechanics* 12: 893–904, 1979.

Millard M, Uchida TK, Seth A, Delp SL. Flexing computational muscle: modeling and simulation of musculotendon dynamics. *Journal of Biomechanical Engineering* 135: 021005, 2013.

Mochon S, McMahon TA. Ballistic walking. *Journal of Biomechanics* 13: 49–57, 1980.

Modica JR, Kram R. Metabolic energy and muscular activity required for leg swing in running. *Journal of Applied Physiology* 98: 2126–2131, 2005.

Mooney LM, Rouse EJ, Herr HM. Autonomous exoskeleton reduces metabolic cost of human walking. *Journal of NeuroEngineering and Rehabilitation* 11: 151, 2014.

Murray WM, Delp SL, Buchanan TS. Variation of muscle moment arms with elbow and forearm position. *Journal of Biomechanics* 28: 513–525, 1995.

Nocedal J, Wright SJ. *Numerical Optimization* (2nd edition). New York, NY: Springer, 2006.

O'Day J, Syrkin-Nikolau J, Anidi C, Kidziński Ł, Delp S, Bronte-Stewart H. The turning and barrier course reveals gait parameters for detecting freezing of gait and measuring the efficacy of deep brain stimulation. *PLoS ONE* 15: e0231984, 2020.

O'Neill MC, Lee L-F, Larson SG, Demes B, Stern JT Jr, Umberger BR. A three-dimensional musculoskeletal model of the chimpanzee (*Pan troglodytes*) pelvis and hind limb. *The Journal of Experimental Biology* 216: 3709–3723, 2013.

Ong CF, Geijtenbeek T, Hicks JL, Delp SL. Predicting gait adaptations due to ankle plantarflexor muscle weakness and contracture using physics-based musculoskeletal simulations. *PLoS Computational Biology* 15: e1006993, 2019.

Ong CF, Hicks JL, Delp SL. Simulation-based design for wearable robotic systems: an optimization framework for enhancing a standing long jump. *IEEE Transactions on Biomedical Engineering* 63: 894–903, 2016.

Pataky TC, Mu T, Bosch K, Rosenbaum D, Goulermas JY. Gait recognition: highly unique dynamic plantar pressure patterns among 104 individuals. *Journal of the Royal Society Interface* 9: 790–800, 2012.

Piazza SJ, Delp SL. The influence of muscles on knee flexion during the swing phase of gait. *Journal of Biomechanics* 29: 723–733, 1996.

Raibert MH, Brown HB Jr, Chepponis M. Experiments in balance with a 3D one-legged hopping machine. *The International Journal of Robotics Research* 3: 75–92, 1984.

Raibert MH, Sutherland IE. Machines that walk. *Scientific American* 248: 44–53, 1983.

Rajagopal A, Dembia CL, DeMers MS, Delp DD, Hicks JL, Delp SL. Full-body musculoskeletal model for muscle-driven simulation of human gait. *IEEE Transactions on Biomedical Engineering* 63: 2068–2079, 2016.

Rajagopal A, Kidziński Ł, McGlaughlin AS, Hicks JL, Delp SL, Schwartz MH. Pre-operative gastrocnemius lengths in gait predict outcomes following gastrocnemius lengthening surgery in children with cerebral palsy. *PLoS ONE* 15: e0233706, 2020.

Ralston HJ. Energy-speed relation and optimal speed during level walking. *European Journal of Applied Physiology* 17: 277–283, 1958.

Rankin JW, Rubenson J, Hutchinson JR. Inferring muscle functional roles of the ostrich pelvic limb during walking and running using computer optimization. *Journal of the Royal Society Interface* 13: 20160035, 2016.

Rathkey JK, Wall-Scheffler CM. People choose to run at their optimal speed. *American Journal of Physical Anthropology* 163: 85–93, 2017.

Reinbolt JA, Fox MD, Arnold AS, Õunpuu S, Delp SL. Importance of preswing rectus femoris activity in stiff-knee gait. *Journal of Biomechanics* 41: 2362–2369, 2008.

Ritchie JM, Wilkie DR. The dynamics of muscular contraction. *The Journal of Physiology* 143: 104–113, 1958.

Saul KR, Hu X, Goehler CM, Vidt ME, Daly M, Velisar A, Murray WM. Benchmarking of dynamic simulation predictions in two software platforms using an upper limb musculoskeletal model. *Computer Methods in Biomechanics and Biomedical Engineering* 18: 1445–1458, 2015.

Seth A, Hicks JL, Uchida TK, Habib A, Dembia CL, Dunne JJ, Ong CF, DeMers MS, Rajagopal A, Millard M, Hamner SR, Arnold EM, Yong JR, Lakshmikanth SK, Sherman MA, Ku JP, Delp SL. OpenSim: Simulating musculoskeletal dynamics and neuromuscular control to study human and animal movement. *PLoS Computational Biology* 14: e1006223, 2018.

Seth A, Matias R, Veloso AP, Delp SL. A biomechanical model of the scapulothoracic joint to accurately capture scapular kinematics during shoulder movements. *PLoS ONE* 11: e0141028, 2016.

Sherman MA, Seth A, Delp SL. What is a moment arm? Calculating muscle effectiveness in biomechanical models using generalized coordinates. *Proceedings of the ASME International Design Engineering Technical Conferences*, Portland, OR, August 4–7, 2013.

Shull PB, Silder A, Shultz R, Dragoo JL, Besier TF, Delp SL, Cutkosky MR. Six-week gait retraining program reduces knee adduction moment, reduces pain, and improves function for individuals with medial compartment knee osteoarthritis. *Journal of Orthopaedic Research* 31: 1020–1025, 2013.

Silder A, Besier T, Delp SL. Running with a load increases leg stiffness. *Journal of Biomechanics* 48: 1003–1008, 2015.

Silverman AK, Neptune RR. Muscle and prosthesis contributions to amputee walking mechanics: a modeling study. *Journal of Biomechanics* 45: 2271–2278, 2012.

Simpson CS, Welker CG, Uhlrich SD, Sketch SM, Jackson RW, Delp SL, Collins SH, Selinger JC, Hawkes EW. Connecting the legs with a spring improves human running economy. *The Journal of Experimental Biology* 222: jeb202895, 2019.

Steele KM, DeMers MS, Schwartz MH, Delp SL. Compressive tibiofemoral force during crouch gait. *Gait & Posture* 35: 556–560, 2012.

Steele KM, Seth A, Hicks JL, Schwartz MH, Delp SL. Muscle contributions to vertical and fore-aft accelerations are altered in subjects with crouch gait. *Gait & Posture* 38: 86–91, 2013.

Thelen DG, Anderson FC, Delp SL. Generating dynamic simulations of movement using computed muscle control. *Journal of Biomechanics* 36: 321–328, 2003.

Thompson JA, Tran AA, Gatewood CT, Shultz R, Silder A, Delp SL, Dragoo JL. Biomechanical effects of an injury prevention program in preadolescent female soccer athletes. *The American Journal of Sports Medicine* 45: 294–301, 2016.

Thomson W. *Popular Lectures and Addresses*. Volume I: *Constitution of Matter*. London: Macmillan & Co., 1889.

TMR. Digital health market: global industry analysis, size, share, growth, trends and forecast, 2017–2025. Transparency Market Research, Report No. TMRGL12473, 2017.

Uchida TK, Hicks JL, Dembia CL, Delp SL. Stretching your energetic budget: how tendon compliance affects the metabolic cost of running. *PLoS ONE* 11: e0150378, 2016a.

Uchida TK, Seth A, Pouya S, Dembia CL, Hicks JL, Delp SL. Simulating ideal assistive devices to reduce the metabolic cost of running. *PLoS ONE* 11: e0163417, 2016b.

Uhlrich SD, Silder A, Beaupre GS, Shull PB, Delp SL. Subject-specific toe-in or toe-out gait modifications reduce the larger knee adduction moment peak more than a non-personalized approach. *Journal of Biomechanics* 66: 103–110, 2018.

Umberger BR, Gerritsen KGM, Martin PE. A model of human muscle energy expenditure. *Computer Methods in Biomechanics and Biomedical Engineering* 6: 99–111, 2003.

Vasavada AN, Li S, Delp SL. Influence of muscle morphometry and moment arms on the moment-generating capacity of human neck muscles. *Spine* 23: 412–422, 1998.

Ward SR, Eng CM, Smallwood LH, Lieber RL. Are current measurements of lower extremity muscle architecture accurate? *Clinical Orthopaedics and Related Research* 467: 1074–1082, 2009.

Weyand PG, Sternlight DB, Bellizzi MJ, Wright S. Faster top running speeds are achieved with greater ground forces not more rapid leg movements. *Journal of Applied Physiology* 89: 1991–1999, 2000.

WHO. Global recommendations on physical activity for health. World Health Organization, 2010.

Wilkie DR. The mechanical properties of muscle. *British Medical Bulletin* 12: 177–182, 1956.

Wingerson L. The lion, the spring and the pendulum. *New Scientist* 97: 236–239, 1983.

Yagn N. Apparatus for facilitating walking, running, and jumping. United States Patent No. 420179, 1890.

Yong JR, Dembia CL, Silder A, Jackson RW, Fredericson M, Delp SL. Foot strike pattern during running alters muscle-tendon dynamics of the gastrocnemius and the soleus. *Scientific Reports* 10: 5872, 2020.

Yong JR, Silder A, Delp SL. Differences in muscle activity between natural forefoot and rearfoot strikers during running. *Journal of Biomechanics* 47: 3593–3597, 2014.

Zajac FE. Muscle and tendon: properties, models, scaling, and application to biomechanics and motor control. *Critical Reviews in Biomedical Engineering* 17: 359–411, 1989.

Zajac FE. Muscle coordination of movement: a perspective. *Journal of Biomechanics* 26: 109–124, 1993.

Zajac FE, Gordon ME. Determining muscle's force and action in multi-articular movement. *Exercise and Sport Sciences Reviews* 17: 187–230, 1989.

Image Credits

All images © 2020 David Delp unless otherwise noted.

Chapter 1

PAGE XX   Vitor Luiz as The Creature in Liam Scarlett's *Frankenstein*. Photo © 2016 Alastair Muir.

PAGE 3   Leonardo da Vinci drawing from Royal Collection Trust, © Her Majesty Queen Elizabeth II 2019, https://www.rct.uk/collection/919003.

PAGE 6   Eadweard Muybridge, "Jumping; Running Straight High Jump," Plate 156 from *Animal Locomotion*, 1887, Volume VII, "Males & Females Draped & Misc. Subjects, 1885." Addison Gallery of American Art, Phillips Academy, Andover, MA / Art Resource, NY.

PAGE 6   Photo of a jumping lizard by Thomas Libby, Evan Chang-Siu, and Pauline Jennings, Courtesy of PolyPEDAL Laboratory and CiBER, University of California, Berkeley.

PAGE 8   *Avatar*, © 2009 Twentieth Century Fox. All rights reserved.

PAGE 12   Data from the *Open Motion Project*, Advanced Computing Center for the Arts and Design, The Ohio State University.

Chapter 2

PAGE 24   A toddler's first steps. Photo by David Scofield by Pexels.

Chapter 3

PAGE 54   Running on a prosthetic limb, Amy Palmiero-Winters set a world record at the 2006 Chicago Marathon. Photo © Matt Furman.

Chapter 4

PAGE 80   Image of skeletal muscle showing the striated microstructure. This is the first *in vivo* image of individual sarcomeres in a live human. Courtesy of Mike Llewellyn.

364 IMAGE CREDITS

Chapter 5

PAGE 104 Thinking about the math behind the muscle.

Chapter 6

PAGE 132 Detail of an illustration from *De motu animalium* (On Animal Motion, 1680) by Giovanni Borelli.

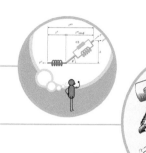

Chapter 7

PAGE 160 An elephant runs while high-speed cameras capture the changing positions of the dots painted on its skin. Photo © 2003 Richard C. Lair.

PAGE 163 Eadweard J. Muybridge, "'Lizzie M.' Trotting, Harnessed to Sulky," 1884–1886, Plate 609 from *Animal Locomotion* (1887). Photo courtesy of Addison Gallery of American Art, Phillips Academy, Andover, MA / Art Resource, NY.

PAGE 164 Fluoroscopy images of the motion of a healthy shoulder and a reverse total shoulder arthroplasty provided by Heath Henninger, Christopher Kolz, and Hema Sulkar from the University of Utah.

Chapter 8

PAGE 192 Acrobats perform an act worthy of reverse engineering at Absinthe in Las Vegas. Photo by Steven Lawton.

Chapter 9

PAGE 216 The covariance matrix adaptation evolution strategy algorithm homing in on a solution.

Chapter 10

PAGE 248 Human gait simulation (center) from Rajagopal et al. (2016). Chimpanzee locomotion simulation (left) from O'Neill et al. (2013). Human gait simulation augmented with an exoskeleton (right) provided by Matthew O'Neill and Brian Umberger.

Chapter 11

PAGE 272 User of an exoskeleton rises from his chair and begins walking. Image courtesy of Ekso Bionics.

Chapter 12

PAGE 304 Three-dimensional muscle-driven simulation of running showing muscle activations.

Chapter 13

PAGE 330 Jacqueline Joyner-Kersee, an Olympic gold medalist in the heptathlon and the long jump. Photo by Tony Duffy.

Index

A-bands, 87–88
Abduction, 19, 186–187
Abduction, hip, 19, 106, 145, 302
Accelerometers. *See* Inertial measurement units (IMUs)
Acetylcholine, 94
Achilles tendon, 21, 106, 144, 306
  foot-strike patterns in running, 319–320, 322
  hopping, 63
  walk-to-run transition, 306, 315–317
Actin, 83–87, 93
Action potentials, 95
Activation dynamics, 100–101, 124, 156–157. *See also* Muscle activation
Active force–length curve, 88–90, 107, 122–123
Activity inequality, 336–337
Adduction, 19, 186–187
Adduction, hip, 19, 106
Adenosine diphosphate (ADP), 86, 92
Adenosine triphosphate (ATP), 85–86, 92, 113
Agonist muscles, 148–149
Anatomical landmarks, 20, 172–174
Anatomical planes and directions, 18
Anatomy, lower limb, 20–21
Anatomy, upper limb, 3, 187

Angles of rotation, 175–176
Angular momentum, 6, 43, 195–196, 317–318
Animal locomotion, 4, 61, 70
  cheetah, 305
  chicken, 114
  elephant, 36, 60–61, 160
  horse, 73, 163, 305
  kangaroo, 61–64
  lizard, 6
  ostrich, 344–345
  Tyrannosaurus rex, 55, 61
Ankle
  flexion in running, 75–76, 144, 211
  flexion in walking, 47–48, 210
  injuries, 345–346
  motions, 18
  muscle moment arm, 138
  planar ankle model, 238
  static optimization problem, 223–226
Ankle dorsiflexion, 18, 44–45
Ankle plantarflexors, 134–135, 144
  assistive device, 300–303
  and center-of-mass acceleration, 308–310
  crouch gait, 293, 296
  dynamic coupling, 251–253
  joint moments, 209, 223–225, 227–228, 235, 237

  running, 314–316
  weakness of, 297–299
Antagonist muscles, 148–149
Anterior, 18–19
Anterior cruciate ligament (ACL), 161, 188–191
Aponeurosis, 85, 120
Archimedes, 133, 135
Aristotle, 4
Arm swing, 42–43, 317–318
Assistive devices, 242–243
  ankle brace, 280, 297
  jumping, 243–245
  running, 300, 323–329
  walking, 299–303
Astronauts, 25–26, 36
Atlas robot, 65
Atypical gait, 49–52
Australopithecus afarensis, 270

Balance, 28, 31
Ballistic walking model, 33–34, 37–39, 281
Biarticular muscles, 252–253
Biomechanical models. *See also* Musculoskeletal models; Simulations
  ballistic walking model, 33–34, 37–39, 281
  bipedal mass–spring model, 74–75
  data collection, 11–15

dynamic walking model, 39–42
inverted pendulum model, 34–36, 38, 74
limitations of, 44, 199–201
mass–spring models, 62–63, 66–67, 74–75, 306
muscle activation, 100–103
muscle force production, 106
planar ankle model, 238
planar standing long jump model, 243–245, 264–265
skeletal model for gait analysis, 44–45
Zajac's model, 106
Biomechanics of movement, 2, 331
benefits of, 7
history of, 4–6
and machine learning, 338–339
and other disciplines, 6–7
terminology, 16–22
Bipedal mass–spring model, 74–75
Bipennate muscles, 109
Body mass, 70–71
Body segments (lower limb), 21
Bones
lower limb, 20
and muscle moment arm, 134
shoulder, 187
Borelli, Giovanni, 4–5
Bottom-up inverse dynamics algorithm, 208
Botulinum toxin, 283, 285–286
Brain surgery, 332

Cadaveric measurement, 119–121
Cadence, 27, 57
Calcaneal gait, 297–299
Calcaneus, 20, 106
Candidate solutions, 221
Cartilage, 193, 213–214
Center-of-mass acceleration, 28–29, 276–280, 289, 291–294, 308–310
Center of pressure, 197–199
Central nervous system (CNS), 94–95, 98, 113

Centripetal force, 35
Cerebellar ataxia, 28
Cerebral palsy, 49–50, 133, 146, 273–274, 339–341
Compliant tendons, 116–117, 126–128
Computed muscle control algorithm, 261
Computer-generated imagery, 7–8
Computer simulations, 13–15, 340–342. *See also* Simulations
Concentric contractions, 92
Constrained inverse kinematics, 183–186
Constraints, 221
Contraction dynamics, 100, 127, 156–157, 269
Contractions, concentric/eccentric, 92
Contraction velocity, 92, 108, 112–114, 124
Contracture, 297–299
Cost function, 221
Cost of transport. *See also* Energy consumption
and gait transitions, 73–74
kangaroo, 61–64
running, 73–74
walking, 37–39, 41, 301–302
Covariance matrix adaptation evolution strategy (CMA-ES), 229–230, 244
Cross-bridge cycle, 85–86, 88, 91
Cross-bridges, 84
Crouch gait, 49–50, 52, 133, 273–274
ankle brace, 280, 297
ankle plantarflexors, 293, 296
gastrocnemius, 293–294
and hamstring length, 146–148
knee flexion, 295–296
knee loads, 265, 267
muscle-driven simulations, 292–297
Cybathlon, 1–2, 8–11
Cycling, 8–11, 81–82, 93

Data collection, 11–15
Deep brain stimulation, 7, 332
Defibrillators, 1
Design variables, 221
Dimensionless curves, 122–124
Direct collocation, 342
Distal, 18
Dorsiflexion, ankle, 18, 44–45
Double-support phases in walking, 26–27, 29–31, 38
Dynamic coupling, 251–253
Dynamic optimization, 239–245, 260–262, 297–299
Dynamic simulations. *See also* Inverse dynamics; Muscle-driven simulations
forward dynamics, 14, 194, 240, 271
muscle–tendon, 100, 121–123
Dynamic walking machine, 39, 42
Dynamic walking model, 39–42
Dynamometer, 149–150

Eccentric contractions, 92
Einstein, Albert, 200
Elastic energy storage
hopping, 61–64
running, 69, 306, 317
shoe, 69–70
Elbow
flexor muscles, 142, 149–150
motions, 19
tendon excursions, 142
Electrical stimulation of muscles, 1–2, 5, 9–10, 95, 98
Electromyography (EMG), 12, 98–100, 156–157
and body motions, 250
running, 308–312
simulated muscle activations, 265–266
walking, 228, 278–279, 287–288
Electrophysiology, 5
Elementary rotation matrices, 179–180
Elephants, 36, 60–61, 160

INDEX

Energy consumption, 12, 227
  cost of transport, 37–39, 41, 73–74, 301–302
  cross-bridge cycle, 85–86
  hopping, 61–64
  muscle-driven simulations, 265–268
  regeneration, 92–93
  running, 306, 317, 323–329
  step-to-step transition cost, 41
Epicondyle, 20, 173
Epidural stimulation, 7
Equinus gait, 297–299, 318–319
Essential tremor, 332
Euler, Leonhard, 175, 195–196
Euler angles, 175
Excitation, muscle, 100–101
Exoskeleton devices, 299–303
Extension, 18–19

Fascicle, muscle, 83–85, 108
Fast-twitch muscle fibers, 113–114
Fatigue, 9, 98, 113
Feasible solutions, 221
Femur, 20, 188, 193
Fiber force, 96
Fibers. *See* Muscle fibers
Fiber velocity, 316
Filmmaking, 6–8
Finite-element models, 129–130
Flexion, 18–19
Flexor digitorum superficialis, 106
Flight phase, 57, 61
Fluoroscopy, 163–165
Foot progression angle, 28, 213–214
Foot slap, 287, 289
Foot-strike patterns, 318–323
Force–deformation curves, 70
Force enhancement, 128
Force–length curves, 107–108, 112, 115, 117, 122–123
Force–length relationships, 88–90, 107
Force–length–velocity–activation relationship, 102–103

Force plates, 12, 28, 196–197
Force–velocity curves, 90–91, 107–108, 112, 122–123
Force–velocity relationship, 90–93, 107
Forensic engineers, 193–194
Forward dynamic simulations, 14, 194, 240, 271
Forward kinematics, 171
Forward progression, 277, 280
Fosbury, Dick, 217
Fosbury flop, 217–219, 242
Free-body diagrams, 135, 202, 205–207
Free moment, 197–198
Freezing-of-gait episodes, 165–166
Frontal plane, 18–19
Froude number, 34–36, 38
Functional electrical stimulation, 1–2, 5, 95
  implanted electrodes, 9, 99
  and muscle fiber recruitment, 10, 98
Fused tetanus, 96

Gaits. *See also* Crouch gait
  atypical, 49–52
  calcaneal/equinus, 297–299, 318–319
  freezing-of-gait episodes, 165–166
  gait transitions, 72–74, 306, 314–317
  heel-walking/toe-walking, 297–299, 318–319
  pentapedal, 62
  retraining, 212–214
  running gait cycle, 56–58, 211, 305–306
  skeletal model for analysis of, 44–45
  sprinting, 144, 314–315
  stiff-knee, 50–52, 282–287
  toe-in, 212–214
  walking gait cycle, 26–28, 148–149, 210
Galvani, Luigi, 1, 5

Gastrocnemius, 20–21, 106, 113, 224–225, 238. *See also* Ankle plantarflexors
  assisted running, 324–325
  assisted walking, 302–303
  and center-of-mass acceleration, 279–280, 309–310
  crouch gait, 293–294
  dynamic coupling, 252–253
  foot-strike patterns in running, 320–323
  gastrocnemius lengthening surgery, 339–341
  stance phase of running, 233, 308–309
  stance phase of walking, 279–281
  and walking speed, 287–291
  and walk-to-run transition, 315–317
Genetic algorithms, 228
Gimbal lock, 181
Gini index, 336
Gluteus maximus, 20, 130, 145, 154–155, 291
  stance phase of running, 233, 308–310
  stance phase of walking, 277–280
Gluteus medius, 20–21, 106, 291, 296
  stance phase of running, 308
  stance phase of walking, 277–280
Gracilis, 106
Gravity, 25–26, 32, 36
  and dynamic walking, 39–42
  and running, 59–60
Greater trochanter, 20, 173
Groucho walk, 36, 60–61
Ground reaction forces
  force plates, 12, 28, 196–197
  inverse dynamics, 194–195, 201–207
  measuring, 196–197
  and muscle force generation, 250
  running, 55, 57–61, 71–73, 77–78, 310, 314–315
  walking, 28–32, 48–49, 72–73, 275–280

Hamstrings, 21, 133
    and crouch gait, 146–148
    and hip flexion, 253
    swing phase of walking, 281–283
    and walking speed, 287–288
Hand tremor, 332–334
Head, arms, and torso (HAT) model, 202, 207–208
Head Injury Criterion (HIC), 166
Heel-walking, 297–299
Henneman's size principle, 98, 113
High-jump technique, 217–219
Hill, A. V., 82, 90, 92, 106, 121–122
Hill-type model of muscle–tendon dynamics, 121–129
Hip
    abduction, 106, 145, 302
    motions, 19, 44–45
Hip flexion, 19
    and hamstrings, 253
    in running, 75–76, 325–326
    skeletal model, 44–45
    in walking, 46–48
Hippocrates, 342
Hopping, 61–64
Hopping robots, 64–65
Horse in Motion photographs, 163
Horses, 73, 163, 305
Human Walking, 149
Huxley, Andrew, 84, 88, 129
Huxley, Hugh, 83–84, 88
Huxley-type model, 129

I-bands, 87–88
Iliopsoas, 20, 154–155, 233, 281–282
Iliotibial band, 285–286
Imaging, 12–13
Impaired movement, 335–336
Implanted electrodes, 9
Inertial frame, 177–178
Inertial measurement units (IMUs), 12–13, 164–166, 332–338
Inferior, 18
Injury prevention, 166, 189–190, 323, 345–346
Inman, Verne, 149

Internal model of body, 220
Intersegmental forces, 251–252
Inverse dynamics, 14–15, 156–157, 193–215, 271
    algorithms, 199–201
    center of pressure, 197–199
    dynamic consistency, 208–209
    external forces, 195–197
    with ground reaction forces, 194–195, 201–207
    without ground reaction forces, 207–208
    and inverse kinematics, 194
    joint moments, 209–212
    verification, 208–209
Inverse kinematics, 14–15, 162
    constrained, 183–186
    and inverse dynamics, 194
    unconstrained, 171–174, 181–183
    underlying skeletal model, 44–45, 183–184
    valgus moments, 190
Inverted pendulum model, 34–36, 38, 74, 135
Isometric contraction, 91–92

Joint accelerations, 251–253
Joint angles. *See also* Kinematics
    computing, 170–174, 181–186, 199
    running speed, 314
Joint loads, 193–195, 214, 238–239
Joint moments
    ankle plantarflexors, 134–135, 209, 223–225, 227–228
    and joint loads, 214
    maximum, 148–152
    running and walking, 209–212, 231, 235, 237, 256
Joint motions, 18–19
Joint movements
    angular velocity, 143–144
    multi-joint muscles, 145–148, 252–253
Jumping, 6, 217–219, 242–245, 264–265

Kangaroos, 61–64
Kelvin, William Thompson, Baron, 162
Kinematics
    forward/direct, 171
    inverse, 14–15, 162, 171–174, 181–186, 190, 194
    running, 75–76
    shoulder model, 186–188
    underlying skeletal model, 44–45, 183–184
    walking, 46–48
Knee
    anterior cruciate ligament (ACL), 161, 188–191
    crouch gait loads, 265, 267
    gait retraining, 212–214
    model complexity, 172, 201
    motions, 18
    musculoskeletal model of, 263–264
    osteoarthritis, 193, 212–214
    valgus position, 189–190
Knee extension, 18, 44–45, 201, 252–253, 281–282
Knee flexion, 18
    angle measurement, 172–174, 181–182
    crouch gait, 49–50, 295–296
    joint acceleration, 251–252
    and rectus femoris, 283–286
    in running, 75–76
    stiff-knee gait, 50–51, 282–287
    and swing phase, 280–282
    in walking, 47–48
    walking speed, 289–292

Lab frame, 177–178
Landing, 189–190, 345
Large-scale studies, 12–13, 338
Laser diffraction, 120
Lateral, 18
Leg stiffness, 70–71
Leonardo da Vinci, 3–4
Ligament, 161, 188–191
Line of progression, 27
List, pelvic, 19, 45–46

Lizard, 6
Locomotion, 17. *See also* Animal locomotion; Movement
Long jump model, 243–245, 264–265
Lower-extremity model, 224, 233, 257
Lumbar spine, 20, 346

Machine learning, 338–339
Magnetic resonance imaging (MRI), 138
Malleolus, 20, 173
Man versus Horse Marathon, 305
Marey, Étienne-Jules, 196
Marker registration, 185
Mass–spring models, 62–63, 66–67, 74–75, 306
Maximum contraction velocity, 92, 108, 112–114, 124
Maximum isometric moment, 150
Maximum isometric muscle force, 111–112, 121, 124, 128
McMahon, Thomas, 33–34, 66–69
Mechanical advantage, 134–136. *See also* Muscle moment arms
Medial, 18
Metabolic cost. *See* Energy consumption
Meta-optimization, 244
Microendoscopy, 80, 120–121
M-lines/-discs, 87–88
Mobility limitations, 335
Models. *See* Biomechanical models; Musculoskeletal models
Moment arm. *See* Muscle moment arms
Motion capture, 6–8, 11–12, 163, 166–171
Motion sensing technology, 332–337
Motivation, 342–343
Motor impairments, 98
Motor neuron, 1, 94, 97–98
Motor unit recruitment, 96–98, 113

Movement, 4–7. *See also* Muscle-driven simulations
 analysis of, 17, 161–162
 and health, 4
 impaired, 335–336
 measurement techniques, 162–171
 motivating, 342–343
 production of, 17
 simulating, 259–262
 terminology of, 16–22
Multipennate muscles, 109
Muscle. *See also* Muscle biology; Muscle fibers; Muscle force generation; Muscle force optimization; Muscle moment arms
 agonist/antagonist, 148–149
 biarticular, 252–253
 bipennate/multipennate, 109
 contractions, 92
 electrical stimulation of, 1–2, 5, 9–10, 95, 98
 excitation, 100–101, 259–260, 285
 fascicle, 83–85, 108
 force enhancement, 128
 and ground reaction forces, 28
 Hill-type model of muscle–tendon dynamics, 121–123
 lower-extremity parameters, 118–119
 mechanical power, 91, 93
 motor units, 97–98
 movement of, 4–5
 neuromuscular stimulators, 9
 nomenclature, 20–21
 parallel-fibered, 109
 pennate/unipennate, 109
 short-range stiffness, 128
 skeletal, 82–83, 87–88
 specific tension, 111, 121
 striated/smooth, 87–88
 velocity of, 91–92
 walking models, 34, 48, 274
Muscle activation, 93–103
 activation dynamics, 100–101, 124, 156–157

 and assisted running, 324–325
 and EMG patterns, 265–266
 force–length–velocity–activation relationship, 102–103
 muscle-driven simulations, 255
 and running, 310–314
 and walking, 287–288
Muscle architecture, 105–106
 animal, 114
 maximum isometric muscle force, 111–112, 121, 124, 128
 maximum muscle contraction velocity, 92, 108, 112–114, 124
 and moment arms, 152–153
 muscle fiber pennation angle, 109–111, 120
 muscle-specific parameters, 117–121
 optimal muscle fiber length, 107–109, 120, 124
Muscle biology, 81–103
 cross-bridge cycle, 85–86, 88, 91
 force–length relationships, 88–90, 107
 force–length–velocity–activation relationship, 102–103
 force–velocity relationship, 90–93
 mechanical work, 82, 92
 motor unit recruitment, 96–98, 113
 muscle activation, 93–103, 124, 156–157
 muscle structure, 83–85
 rate encoding, 95–96
 sarcomeres, 80, 83–84, 86–88
 skeletal muscle, 82–83, 87–88
Muscle-driven running, 305–329
 building and testing simulations, 307–310
 device-assisted running, 323–329
 foot-strike patterns, 318–323
 leg swing, 317–319
 and mass–spring models, 306
 muscle activity, 308–313
 muscle–tendon dynamics, 317
 running gait, 305–306
 running speed, 312–314

370 INDEX

run-to-sprint transition, 314–315
spring-enhanced running, 326–329
walk-to-run transition, 306, 315–317
Muscle-driven simulations, 2, 105, 128, 249–271. *See also* Muscle-driven running; Muscle-driven walking
   analyzing, 269–270
   creating, 254
   dynamic coupling, 251–253
   energy consumption, 265–268
   and experimental measurements, 264–265
   modeling musculoskeletal dynamics, 255–259, 275
   optimization strategies, 260–262
   sensitivity, 268
   simulating movement, 259–262
   validation, 227–228, 242, 244–245, 262–268
   verification, 267–268
Muscle-driven walking, 261–262, 273–303
   assistive devices, 299–303
   building and testing simulations, 275
   center-of-mass acceleration, 276, 280, 289, 291–294
   crouch gait, 292–297
   forward progression, 277, 280
   future effects of muscle activity, 285
   ground reaction forces, 275–280
   heel-walking/toe-walking, 297–299, 318–319
   loaded and inclined walking, 261–262, 268
   muscle activity, 276–279, 287–292
   stiff-knee gait, 50–52, 282–287
   swing phase, 253, 280–282
   walking speeds, 287–292
Muscle fibers, 83–85, 94, 97
   contraction velocity, 92, 108, 112–114, 124

fast-twitch/slow-twitch, 113–114
force and velocity, 96, 316
optimal length, 107–109, 120, 124
pennation angle, 109–111, 120
recruitment, 10, 96–98, 113
twitch, 1, 95–96
types of, 113
Muscle force generation, 84, 86, 102–103, 106–107
   compliant tendon, 116–117, 126–128
   force–length relationships, 88–90, 107
   force–velocity relationship, 90–93, 107
   and ground reaction forces, 250–251
   Hill-type model, 121–129
   and joint loads, 214, 238–239
   and moment arm, 143–145
   other models, 128–130
   rigid tendon, 116–117, 124–126, 262–263
   running, 232, 236
   walking, 234
Muscle force optimization, 217–245
   biological and numerical optimizers, 220–223
   dynamic optimization, 239–245
   generic optimization problem, 221–222
   global methods, 228–230
   joint loads, 238–239
   local methods, 226–228
   muscle coordination, 219, 242–245
   muscle redundancy problem, 219, 221, 223, 230
   static optimization problems, 223–230, 239
   walking and running, 230–237
Muscle-induced acceleration analysis, 269

Muscle moment arms, 134–148
   ankle plantarflexion moments, 223–225, 227–228
   defined, 137–138
   and joint angular velocity, 143–144
   in lower limb, 224, 233
   maximum joint moments, 148–152
   measuring, 140–142
   multi-joint muscles, 145–148
   and muscle architecture, 152–153
   and muscle force generation, 143–145
   and muscle lengths/velocities, 143–145
   muscles with complex actions, 154–156
   tendon-excursion definition of, 138–142
   in upper limb, 142
Muscle redundancy/force-sharing problem, 219, 221, 223, 230
Muscle–tendon actuator, 122
Muscle–tendon model, 100, 110, 121–123, 128–129
Musculoskeletal geometry, 133–157. *See also* Muscle moment arms
   maximum joint moments, 148–152
   moment arm definitions, 137–142
   multi-joint muscles, 145–148
   muscle length and velocity, 143–145
   muscles with complex actions, 154–156
   standing posture, 134–136
   tendon transfer surgery, 152–153
Musculoskeletal models, 14. *See also* Biomechanical models; Muscle-driven simulations
   activation/contraction dynamics, 156–157
   complex muscles, 155–156
   cycling, 9
   degrees of freedom, 201, 256–257

finite-element models, 129–130
and hamstring lengthening surgery, 146–148
Hill-type model of muscle–tendon dynamics, 121–123, 128–129
Huxley-type model, 129
inverted pendulum model, 135
knee model, 263–264
lower extremity, 224, 233, 257
maximum joint moments, 148–152
measuring parameters, 119–121
muscle-driven simulations, 255–259, 275
muscle–tendon units, 137
neck, 258, 346
planar upper/lower extremity, 256–257
running, 307, 346
scaling, 185, 257–259
selection of, 199–201, 256–257
shank and foot model, 224
shoulder kinematics, 186–188
squat model, 199–207
underlying skeletal model, 44–45, 183–184
upper extremity, 256–259, 346
validation, 141–142, 188, 262–268
via points/wrapping surfaces, 155
Muybridge, Eadweard, 5–6, 163, 196
Myocytes, 84. *See also* Muscle fibers
Myofibrils, 83–84, 87–88
Myosin, 83–87

Neck models, 258, 346
Nervous system, 2, 96
Neuromuscular control, 340–341
Neuromuscular junction, 94, 97
Neuromuscular stimulators, 9
Newton, Sir Isaac, 193, 195
  second law of motion, 28, 195–196, 203, 205
  third law of motion, 202–203, 218
Numerical optimization, 220–223

Obesity, 336–337
Objective function, 221
"On the Method of Theoretical Physics," 200
On the Motion of Animals, 4
Open science, 343–346
OpenSim, 148, 190, 201–202, 209, 267–268, 270–271, 345–346
OpenSim models, 270–271. *See also* Musculoskeletal models
Optical motion capture, 7–8, 11–12, 166–171
Optimal muscle fiber length, 107–109, 120, 124
Optimal sarcomere length, 88–89, 107
Optimal solution, 221
Optimal tracking problem, 260–261
Optimization, 183–185, 244, 341–342. *See also* Muscle force optimization
  dynamic, 239–242, 260–262
  local/global optimum, 219, 226, 228
  numerical, 220–223, 228–230
  problem statements, 222, 224, 232
  static, 223–230, 239
  strategies, 260–262
Orderly recruitment, 98
Orthogonality, 177
Orthoses. *See* Assistive devices
Osteoarthritis, 193, 212–214
Ostrich, 344–345

Paleontologists, 55, 134
Parallel-fibered muscles, 109
Paralysis, 1, 7–10, 152, 345
Parkinson's disease, 7, 165–166
Passive force generation, 86–87, 90
Passive force–length relationship, 89–90, 107, 122, 124
Peak isometric stress, 111
Pelvis, 20–21
  motions, 19, 44–45
  in walking, 46

Penalty methods, 221
Pennated muscles, 109–110
Pennation angle, 109–111, 120
Pentapedal gait, 62
Photography, 5–6, 162–163
Physical activity, 4, 13, 336–338, 342–343. *See also* Movement
Physical therapy, 335
Physiological cross-sectional area (PCSA), 111–112, 121
Pitch, 176–177
Planar ankle model, 238
Planar robot, 64–65
Planar standing long jump model, 243–245, 264–265
Planar upper/lower-extremity models, 256–257
Plantarflexion, ankle, 18
Plantarflexors, ankle, 134–135, 144
  assistive device, 300–303
  and center-of-mass acceleration, 308–310
  crouch gait, 293, 296
  dynamic coupling, 251–253
  joint moments, 209, 223–225, 227–228, 235, 237
  running, 314–316
  weakness of, 297–299
Polar bears, 158, 246, 348, 352
Posterior, 18–19
Posterior cruciate ligament (PCL), 188–189
Predictive simulations, 239–245, 265, 267–268, 341–342
Prostheses, 7, 54
Proximal, 18
Psoas, 20, 154–155
"Push-me–pull-you" experiment, 81–82, 90, 92

Quadriceps, 20, 149, 293–296, 308, 310
Quadruped. *See* Animal locomotion
Quantitative measurement, 162
Quaternion, 181

Rate encoding, 95–96
Recruitment, motor unit, 96–98, 113
Rectus femoris, 20, 51, 109, 233, 281–288, 291, 324–325. *See also* Quadriceps
Reference frames, 171–174, 180
Rehabilitation, 11, 335
Reinforcement learning, 341–342
Rigid tendons, 116–117, 124–126, 262–263
Robots, 7
   dynamic walking, 39
   hopping, 64–65
   running, 65
Roll, 176–177
Root-mean-squared error (RMSE), 188
Rotation, hip, 19, 44–45
Rotation, pelvic, 19, 45–46
Rotation matrices, 175–181
Royal Society of London, 81
Running, 55–78. *See also* Gaits; Muscle-driven running
   ankle flexion in, 75–76, 144
   arm swing, 317–318
   assistive devices, 300, 323–329
   cost of transport, 73–74
   electromyography, 308–312
   energy consumption, 306, 317, 323–324, 327
   flight phase, 57
   forefoot/rearfoot striking, 58–59, 66, 318–323
   gait cycle, 56–58, 211, 305
   gait transitions, 72–74, 306, 314–317
   ground reaction forces, 55, 57–61, 66, 72–73, 77–78, 310, 314–315
   hip flexion, 75–76, 325–326
   and hopping, 61–64
   joint moments, 209–212, 231, 237
   kinematics of, 75–76
   kinetic/potential energy in, 59–60
   leg stiffness and body mass, 70–71
   loaded, 70–71
   muscle force generation, 232, 236
   muscle redundancy/force-sharing problem, 230–237
   musculoskeletal models, 307, 346
   shoe design, 69–71
   speed, 57, 77–78, 312–314
   swing phase, 56
   tendons in, 317
   track stiffness, 66–68
   and walking, 36, 57, 60, 72–75
Run-to-sprint transition, 314–315

Sacrum, 20
Sagittal plane, 18–19
Sarcolemma, 94
Sarcomeres, 80, 83–84, 86–88, 91, 107, 120
Sarcoplasmic reticulum, 84, 94–95
Sartorius, 20, 109, 285–286
Scapula, 186–188
Semimembranosis, 21, 145–146. *See also* Hamstrings
Sensitivity, 268
Sex-related differences, 189–190, 336
Short-range stiffness, 128
Shoulders
   fluoroscopic images of, 164
   kinematic model of, 186–188
   motions, 19
SIMM: Software for Interactive Musculoskeletal Modeling, 270
simtk.org, 343
Simulated annealing, 230
Simulations. *See also* Biomechanical models; Inverse dynamics; Muscle-driven simulations; Musculoskeletal models
   complexity of, 199–201
   computer, 13–14, 340–342
   forward and inverse, 14–15, 194, 240, 271
   predictive, 239–245, 341
   and surgery, 133–134
Single-support phases in walking, 26–27
Skating equipment, 9
Skeletal model for gait analysis, 44–45
Skeletal muscles, 82–83, 87–88
Skull motion, 166
Sliding filament theory, 88
Slow-twitch muscle fibers, 113–114
Smartphone technology, 12–13, 335–338, 342–343
Soft-tissue artifacts, 168–169, 186, 188
Soleus, 20–21, 113, 224–225, 238, 278–282. *See also* Ankle plantarflexors
   assisted running, 324–325
   assisted walking, 302–303
   and center-of-mass acceleration, 278, 280, 308, 310
   dynamic coupling, 251–252
   foot-strike patterns in running, 320–322
   and running speed, 312
   stance phase of running, 233, 308
   and walking speed, 287–291
   and walk-to-run transition, 315–317
Solution space, 221
Speed
   running, 57, 77–78, 312–314
   walking, 27, 35–36, 38, 48, 56, 287–292
Spot robot, 65
Spring-enhanced running, 326–329
Sprinting, 144, 314–315
Squat model, 199–207
Stance phase, 26–27, 56–57, 277–280, 308–309
Standing long jump, 242–245, 264–265
Standing posture, 134–136

Stanford, Leland and Jane, 162–163, 196
Static calibration trial, 182–183
Static optimization problems, 223–230, 239
Statistical methods, 339
Steepest-descent algorithm, 226–227
Steno, Nicolas, 4–5
Step frequency
　hopping, 62–63
　running, 57
　walking, 27, 41
Step length
　hopping, 62–63
　running, 57, 69
　walking, 27, 41
Step-to-step transition cost, 41
Step width, 27–28
Stiff-knee gait, 50–52, 282–287
Striated muscles, 87–88
Stride frequency, 27, 57, 62–63, 314–315
Stride length
　running, 57, 314–315
　walking, 27
Superior, 18
Surgery
　brain, 332
　gastrocnemius lengthening, 339–341
　hamstring lengthening, 133, 146–148
　and motion measurement, 334
　rectus femoris, 283, 285–287
　and simulations, 133–134
　tendon transfer, 152–153
Swing phase, 26–27, 56, 253, 280–283
Synapse, 94

Tail, 6, 62, 64
Talus, 238
Tendon-excursion definition of moment arm, 138–142

Tendons
　aponeurosis, 85, 120
　compliant, 116–117, 126–128
　force–length curve, 107, 115–116, 122, 124
　free tendon, 85
　Hill-type model of muscle–tendon dynamics, 121–123
　muscle force generation, 116–117, 124–128
　regions of, 115–116
　rigid, 116–117, 124–126, 262–263
　running, 317
　slack length, 114–117, 120–121, 124
Tendon transfer surgery, 152–153
Terminal cisternae, 94–95
Terminology of movement, 16–22
Tetanic frequency, 96
Tibia, 20, 188, 193
Tibialis anterior, 20, 121, 148–149, 224, 233, 287–289
Tibiofemoral contact forces, 267
Tilt, pelvic, 19
Titin, 86, 89–90
Toe-in gait, 212–214
Toe-in/toe-out angles, 27–28
Toe-walking, 297–299, 318–319
Top-down inverse dynamics algorithm, 208
Torque actuator, 129, 243
Track stiffness, 66–68
Transformation matrices, 174–182
Transverse plane, 18–19
Treadmills, 12
Treatment planning, 188, 212–214, 273–274, 296–297, 299, 335, 339–341
Tropomyosin, 86, 93
Troponin, 86, 93
T-tubules, 94
Tuned tracks, 66–69
Type I/II muscle fibers, 113
Tyrannosaurus rex, 55, 61

Ultrasonography, 120–121
Unconstrained inverse kinematics, 171–174, 181–183
Unfused tetanus, 96
Unipennate muscles, 109
Unit testing, 268
Upper-extremity model, 256–259, 346

Valgus position of knee, 189–190
Validation, 141–142, 188, 227–228, 242, 244–245, 262–268
Vasti, 20–21, 233, 277–282, 289–294. *See also* Quadriceps
Ventral intermediate nucleus, 332
Verification, 208–209, 267–268
Video, 6–8, 11–12, 218, 273–274. *See also* Optical motion capture
Virtual work, principle of, 139–140

Walking, 25–53. *See also* Gaits; Muscle-driven walking
　agonist/antagonist muscles, 148–149
　arm swing, 42–43
　assistive devices, 299–303
　ballistic model, 33–34, 37–39, 281
　center-of-mass acceleration, 28–29, 276
　children, 35–36
　cost of transport, 37–39, 41, 301
　device-assisted, 299–303
　dynamic walking model, 39–42
　electromyography, 228, 278–279, 287–288
　Froude number, 34–36, 38
　gait cycle, 26–28, 148–149, 210
　and gravity, 25–26, 32, 36
　Groucho, 36, 60–61
　ground reaction forces, 28–32, 48–49, 275–280
　heel-walking/toe-walking, 297–299, 318–319
　inverted pendulum model, 34–36, 38
　joint moments, 209–212, 235
　kinematics of, 46–48

kinetic/potential energy in, 32–33
loaded and inclined, 261–262, 268
muscle force generation, 234
muscle redundancy/force-sharing problem, 230–237
pressure distribution, 198
and running, 36, 57, 60, 72–75
shank angular velocity, 165–166
skeletal model for gait analysis, 44–45
speed, 27, 35–36, 38, 48, 56, 287–292
swing phase, 26–27, 253, 280–283
under various conditions, 52–53
Walk-to-run transition, 306, 315–317
Wearable motion sensors, 164–165, 332–338
Wearable neurostimulator, 332–334
Wilkie, D. R., 121
Wright, Orville and Wilbur, 39

X-ray images, 163–164, 193, 213, 263

Yaw, 176–177

Zajac, Felix, 106, 249–250, 252–253
Z-lines/-discs, 86–88